Objective Resilience

Other Titles of Interest

Objective Resilience: Objective Processes, **MOP 147**, edited by Mohammed M. Ettouney, Ph.D., P.E. (ASCE, 2021). Illustrates some of the objective processes that are used to manage community and asset resilience and provides infrastructure stakeholders with a comprehensive set of practices. (ISBN 978-0-7844-1589-4)

Objective Resilience: Technology, **MOP 148**, edited by Mohammed M. Ettouney, Ph.D., P.E. (ASCE, 2021). Examines the use of different technologies to enhance community and asset resilience and provides a comprehensive set of practices for infrastructure stakeholders. (ISBN 978-0-7844-1590-0)

Objective Resilience: Applications, **MOP 149**, edited by Mohammed M. Ettouney, Ph.D., P.E. (ASCE, 2021). Provides different applications that aim to enhance community and asset resilience from the community viewpoint. (ISBN 978-0-7844-1591-7)

Objective Resilience

Policies and Strategies

Sponsored by the
Objective Resilience Committee of the
Engineering Mechanics Institute of the
American Society of Civil Engineers

Edited by
Mohammed M. Ettouney, Ph.D., P.E.

Published by the American Society of Civil Engineers

Library of Congress Cataloging-in-Publication Data

Names: Engineering Mechanics Institute. Objective Resilience Committee, author. | Ettouney, Mohammed, editor.

Title: Objective resilience. Policies and strategies / sponsored by the Objective Resilience Committee of the Engineering Mechanics Institute of the American Society of Engineers; edited by Mohammed M. Ettouney, Ph.D., P.E.

Description: [Reston, Virginia] : American Society of Civil Engineers [2022] | Series: ASCE manuals and reports on engineering practice ; no. 146 | Part of a four book committee report comprised of: Policies and strategies; Objective processes; Applications; Technology. | Includes bibliographical references and index. | Summary: "MOP 146 examines policies and strategies related to community and asset resilience and provides infrastructure stakeholders with a comprehensive set of practices"-- Provided by publisher.

Identifiers: LCCN 2021040737 | ISBN 9780784415887 (paperback) | ISBN 9780784483749 (pdf)

Subjects: LCSH: Reliability (Engineering) | Civil engineering--Standards. | Natural disasters--Risk assessment. | Emergency management--Government policy. | Engineering--Management. | Organizational resilience.

Classification: LCC TA169 .E545 2022 | DDC 620/.00452--dc23

LC record available at https://lccn.loc.gov/2021040737

Published by American Society of Civil Engineers
1801 Alexander Bell Drive
Reston, Virginia 20191-4382
www.asce.org/bookstore | ascelibrary.org

MANUALS AND REPORTS ON ENGINEERING PRACTICE

(As developed by the ASCE Technical Procedures Committee, July 1930, and revised March 1935, February 1962, and April 1982)

A manual or report in this series consists of an orderly presentation of facts on a particular subject, supplemented by an analysis of limitations and applications of these facts. It contains information useful to the average engineer in his or her everyday work, rather than findings that may be useful only occasionally or rarely. It is not in any sense a "standard," however, nor is it so elementary or so conclusive as to provide a "rule of thumb" for nonengineers.

Furthermore, material in this series, in distinction from a paper (which expresses only one person's observations or opinions), is the work of a committee or group selected to assemble and express information on a specific topic. As often as practicable, the committee is under the direction of one or more of the Technical Divisions and Councils, and the product evolved has been subjected to review by the Executive Committee of the Division or Council. As a step in the process of this review, proposed manuscripts are often brought before the members of the Technical Divisions and Councils for comment, which may serve as the basis for improvement. When published, each manual shows the names of the committees by which it was compiled and indicates clearly the several processes through which it has passed in review, so that its merit may be definitely understood.

In February 1962 (and revised in April 1982), the Board of Direction voted to establish a series titled "Manuals and Reports on Engineering Practice" to include the manuals published and authorized to date, future Manuals of Professional Practice, and Reports on Engineering Practice. All such manual or report material of the Society would have been refereed in a manner approved by the Board Committee on Publications and would be bound, with applicable discussion, in books similar to past manuals. Numbering would be consecutive and would be a continuation of present manual numbers. In some cases of joint committee reports, bypassing of journal publications may be authorized.

A list of available Manuals of Practice can be found at http://www.asce.org/bookstore.

DEDICATION

This objective resilience manual of practice is dedicated to the essential workers who are exposed daily to the dangers of the COVID-19 pandemic. Included among the many groups of workers are the following: healthcare personnel, first responders, public safety officers, correction facility workers, food and agriculture, grocery store workers, teachers, US postal service workers, public transit workers, and many more people who work tirelessly to maintain a sense of normalcy in these unprecedented times.

CONTENTS

BLUE RIBBON PANEL
(In Alphabetical Order)

Joseph Brennan, R.A. (New York), AIA, is an architect and digital practice evangelist who has worked on projects in various capacities for SHoP Architects, Populous, and Gensler. In addition, he is currently an adjunct assistant professor at Columbia University's Graduate School of Architecture, Planning, and Preservation and has taught design technology–focused courses at various institutions in New York. In addition to practice and teaching, Joseph has mentored students at Columbia University, as well as emerging businesses through the New Museum's New Inc. Incubator program.

James Brunetti, P.E., is currently director of operations for Absolute Civil Engineering Solutions in Ft. Lauderdale, Florida. James has more than 35 years of experience in structural mechanics, structural dynamics, design of structures for extreme loadings, fracture mechanics analysis, forensic engineering, and structural failure analysis. He holds a professional degree in engineering mechanics and an M.Sc. degree in structural engineering from Columbia University, as well as a B.Sc. degree in civil engineering from the University of Virginia.

Albert DiBernardo, P.E., ACC, is a consulting engineer. He is the past president of TAMS Consulting, Inc. and the past executive vice president/principal of Thornton Tomasetti, Inc. DiBernardo began his A/E/C career in 1974 and for more than 43 years worked on civil infrastructure projects worldwide, including water resources, airports, bridges, buildings, port facilities, and environmental projects. In the mid-1990s, he began teaching engineering professionals leadership and management and until 2016 served as an adjunct professor in the NYU Tandon School of Engineering graduate program for architects, engineers, and construction professionals. Today, DiBernardo is still serving professionals in the field as a certified life/career and business coach, business advisor, meeting facilitator, mentor, and teacher.

Ketan Dodhia, P.E., specializes in man-made hazards, blast and progressive collapse at Stone Security Engineering. For nearly 20 years as a structural engineer, Ketan has focused on protective design projects and worked on numerous sensitive public and private sector projects. In addition to working all over the World Trade Center sites, including the National September 11th Memorial and Museum, Calatrava PATH Hub Station, and WTC Towers 2, 3, and 4, Ketan has worked on the New York Stock Exchange, the new Smithsonian National Museum of African American History and Culture, and various US embassies worldwide. As a senior project manager at Stone, Ketan is responsible for project management and leadership of the structural design team on blast-resistant design assignments.

Christopher Doyle is a consultant on homeland security matters, an industry in which he has been immersed for 30 years. Doyle led disaster response and recovery operations both in the field and at the headquarters while with the Federal Emergency Management Agency (FEMA). With the creation of the Department of Homeland Security, Doyle went on to help in the stand-up of the Science and Technology Directorate, directing a portfolio of research and development focused on response and recovery, as well as infrastructure protection. In this role, Doyle led several initiatives to promote the notion of resilience by integrating protection from natural and man-made hazards throughout the built environment, with particular emphasis on critical infrastructure. He was presented with the Institute Award from the National Institute of Building Sciences for his leadership in this area.

Henry Green, Hon. AIA, has held several leadership positions in the building community, including serving as executive director of the Bureau of Construction Codes in the Michigan Department of Labor. Henry was a member of the Building Officials and Code Administrators (BOCA) Board of Directors for 10 years, holding the position of president in 1997. Henry was a founding member of the International Code Council Board of Directors, completing a term as a president in 2006. He served as a member of the National Institute of Building Sciences Board of Directors for 8 years, completing a term as chair in 2003 and serving as a president for more than 10 years. Henry also has served on numerous committees for other building industry organizations and is the recipient of numerous awards. Henry was recognized by the United States House of Representatives for his work as "a tireless advocate for building safety and enforcement of codes."

Ahsan Kareem, Ph.D., P.E., Dist.M.ASCE, NAE, is the Robert M. Moran Professor of Engineering and director of the NatHaz Modeling Laboratory at the University of Notre Dame. His work focuses on probabilistic characterization of dynamic load effects owing to wind, waves, and earthquakes on tall buildings, long-span bridges, offshore structures, and other structures via analytical and computational methods and fundamental experiments at laboratory and full scale.

Sarbjeet Singh, Ph.D., P.E., LEED, is a licensed civil engineer with more than 20 years of industry and research experience, knowledge, and expertise in diverse areas of structural engineering/dynamics related to buildings, bridges, railroad, tunnels, and infrastructure. Currently, Sarbjeet is working as Principal Engineer at Metropolitan Transportation Authority (MTA) Construction and Development, New York. His past industry experience includes working with Weidlinger Associates Inc. and AECOM. Sarbjeet executed his postdoctorate research in structural control at the Virginia Polytechnic Institute and State University, USA. Sarbjeet is currently an active member of the ASCE Technical Committee of Infrastructure Systems with the Transportation & Development Institute and is a member of many professional societies, including ASCE, the Earthquake Engineering Research Institute, and the American Institute of Steel Construction.

AUTHORS
(In Alphabetical Order)

Amar A. Chaker, Ph.D., F.AEI, F.EMI, F.ASCE, earned civil engineering degrees from the ENPC in Paris and from the University of Illinois, Urbana-Champaign. He served on the faculty of the USTHB (Algiers), UIUC, and Drexel University. He is now director of the Engineering Mechanics Institute (EMI) of ASCE.

Ryan Colker is vice president of Innovation, International Code Council, and executive director, Alliance for National and Community Resilience, and former vice president, National Institute of Building Sciences. He is a graduate of George Washington University and the University of Florida.

M. Ettouney, Ph.D., P.E., F.AEI, Dist.M.ASCE, has 52 years of consulting experience in many areas, including in very low-to-ultra-high-frequency dynamics and man-made and natural hazards risk and resilience management. Lately, he has been concentrating on the use of game, decision, graph, and probabilistic graph theories (including developing theoretical interactions between these theories) in infrastructure health, progressive collapse, and climate change.

S. Gerasimidis, Ph.D., is an assistant professor of civil engineering at the University of Massachusetts, Amherst. He has published more than 100 technical papers in peer-reviewed international journals and conference proceedings on infrastructure deterioration/collapse, metamaterials, and shell buckling.

Roger J. Grant is executive director, Building Information Management at NIBS and has 30 years of experience in design, construction, and management of the built environment. He earned a BS in construction management and an MBA from Bradley University.

Milagros Nanita-Kennett is the director of research and innovation at the Instituto de Educación Superior en Formación Diplomática y Consular (INESDYC), Ministerio de Relaciones Exteriores (MIREX). Previously, she served as a program manager and senior architect in US DHS S&T as applied to risk and resilience assessment and mitigation. She managed preparedness, response, and recovery for different natural and man-made disasters, environmental sustainability, and climate resilience.

Zackary Kennett, S.E., has over a decade of experience in design and construction administration of buildings, including new construction, renovations, and buildings on existing steel; RC; and masonry structures. He is an expert in interrelationships between asset and community resilience.

Eric Letvin, P.E., Esq., CFM, serves as FEMA's deputy assistant administrator for mitigation. He directs the FEMA's pre- and postdisaster mitigation grant programs that support sustainable, disaster-resilient communities and to avoid or reduce the loss of life, property, and financial impacts of natural hazards.

Chris Mullen, Ph.D., M.ASCE, is an associate professor of civil engineering at the University of Mississippi. Dr. Mullen has 35 years of experience in structural engineering and mechanics and has been active in SEI and EMI. He earned BSCE and MSCE degrees from Rice University and a Ph.D. from Princeton University.

PREFACE

Engineering is a balance between analysis and design. Objectivity forms, mostly, the basis of mathematics and science, which form, mostly, the basis of analysis. Subjectivity forms, mostly, the basis of art, intuition, and imagination, which form, mostly, the basis of design (see Figure 1). Achieving a proper balance between subjectivity and objectivity during

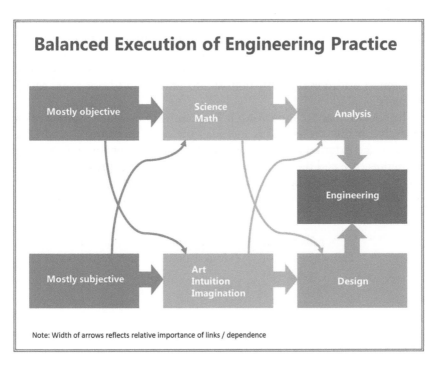

Figure 1. Balanced Execution of Engineering Practice.

the engineering process will ensure an optimal product. This is true especially for complex products that have multitudes of different types of components. Admittedly, community and asset resilience is a complex issue, and as such dealing with it from the engineering viewpoint will require a proper balance between objective and subjective processing.

The Objective Resilience Committee (ORC) of the Engineering Mechanics Institute (EMI) of ASCE was formed in 2015 to help achieve a balanced resilience treatment, especially from an objective viewpoint. Soon after its formation, the ORC initiated the development of an Objective Resilience Manual of Practice (OR-MOP) in 2016. The main objective of the OR-MOP is to provide a comprehensive basis of recommended practices that can help enhance community and asset resilience, while emphasizing the objective side of such practices. The developers of the OR-MOP quickly realized that because of the wide-ranging extent of community and asset resilience, the OR-MOP needed to split its focus into four basic categories: (1) Policies and Strategies, (2) Objective Processes, (3) Technology, and (4) Applications.

This book examines policies and strategies related to community and asset resilience. It aims at providing a comprehensive set of practices, after presenting and discussing the basis for these practices. It is recognized that this OR-MOP is limited, given the limiting factors of space and time, especially in view of the aforementioned wide range extent of resilience. However, the developers hope that the OR-MOP can be used as a guide in developing additional MOPs that would address additional aspects of resilience.

The development of the OR-MOP took almost five years. Many worked tirelessly on this project. This includes the authors of the contributing chapters, the external Blue Ribbon Panel, which independently reviewed the manuscript, and the ASCE Publications editors who provided valuable insights and feedback. Special thanks to Dr. Amar Chaker, the EMI director, for his efforts and help without which this OR-MOP could not have been possible.

Mohammed M. Ettouney, Ph.D., P.E., F.AEI, Dist.M.ASCE
February 2021. West New York, New Jersey

INTRODUCTION

There are several popular definitions for resilience, including NIAC (2009), the NSC (2011), or the Office of the Press Secretary (2013). For example, NIAC (2009), defined infrastructure resilience as follows:

Infrastructure resilience is the ability to reduce the magnitude and/or duration of disruptive events. The effectiveness of a resilient infrastructure or enterprise depends upon its ability to anticipate, absorb, adapt to, and/or rapidly recover from a potentially disruptive event.

As defined, resilience represents a major issue for society, given the magnitude of disaster costs of different kinds. Recognizing the needs of society to build and sustain resilient assets and communities, stakeholders (e.g., federal, state, and local officials, business owners, professionals, educators, and researchers) devoted considerable effort, time, and expense examining asset and community resilience. Given the wide range of factors that affects resilience, knowledge gaps of the subject are still significant. Similar to most important topics, treatment, handling, and communicating resilience-related matters started with a subjective basis. Objective developments lagged their subjective counterparts; however, these developments have been gaining momentum in the last few years. One primary reason for the elevated interest in resilience-related objective processes is that without adequate objectivity, it will remain difficult to provide optimal policies and strategies that aim at delivering practical asset and community resilience at reasonable costs.

Recognizing the needs for comprehensive and practical objective views of asset and community resilience, the Objective Resilience Committee (ORC) of the Engineering Mechanics Institute (MEI) of ASCE embarked on developing an Objective Resilience Manual of Practice (OR-MOP). The

MOPs of ASCE aim at providing discussions, overviews, developments, and/or best practices concerning different topics. To better attain the stated goals, the OR-MOP endeavors to explore and discuss some of the many issues regarding objective resilience. The OR-MOP also strives to provide best practices sections in all the resilience-related subjects it covers. The OR-MOP attempts to address the intersection of three different areas: resilience (*Re*), civil infrastructure (*CI*), and objective processes (*OP*), see Figure 1. In a set-theory formalism, we can express OR-MOP as

$$OR - MOP \equiv Re \cap CI \cap OP \tag{1}$$

Because of the different nature of the chapters of the OR-MOP, we expect that the extent of their treatment of *OP* would vary.

To cast as wide a net for resilience-related objective issues as possible, which is not an easy task in itself, the OR-MOP is subdivided into four books. Each book will examine objective resilience from different viewpoints. Figure 2 illustrates the general subjects of the four books.

This book, Policies and Strategies, is the first book in the OR-MOP collection. The main objective of the book is to explore different policies and

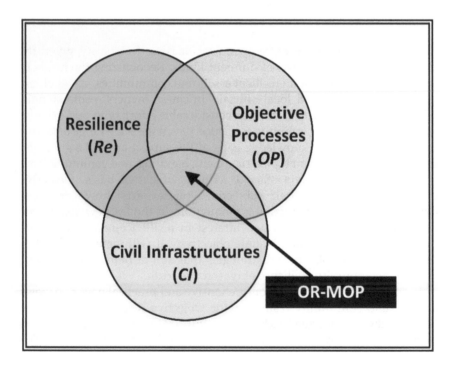

Figure 1. Confluence of domains of the OR-MOP.

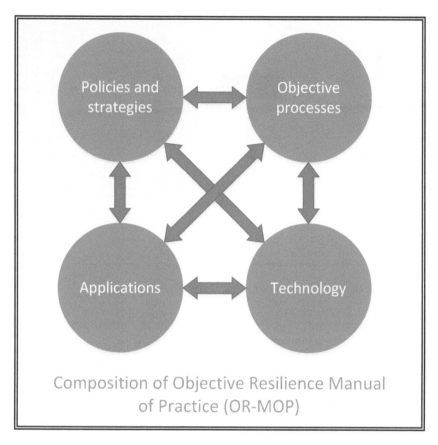

Figure 2. Composition of the OR-MOP.

strategies that can provide for resilient assets and communities. Recognizing that any successful objective process needs to start with the definitions of its subjects, we offer an overview of the many definitions of resilience in the first chapter of the OR-MOP. The chapter also introduces a unifying theory for resilience definitions (TRDs) regarding the interrelationships between the components of some of the popular resilience definitions. The TRD should streamline objective applications of resilience. Chapters 2 through 6 provide different outlooks of policies and strategies, as applied to resilience. Chapters 2 through 4 explore relevant policies and strategies of organizations, codes and standards, and the federal government, respectively. Chapter 5 presents a review of objective modeling considerations from policy/strategy viewpoints. Chapter 6 offers a discussion of resilience management. All chapters will propose a set of recommended practices at the conclusion of each chapter. See Figure 3 for a map of the organization of the book.

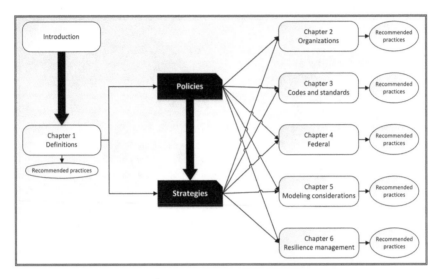

Figure 3. Map of this book (policies and strategies).

The intended readers of this OR-MOP include all civil infrastructure stakeholders, which may broadly include the following:

- Public and private civil infrastructure organizations (transportation, water resources, bridges, health care, and so on);
- City, county, and state officials;
- Emergency managers;
- Public safety personnel;
- Facility managers;
- Security consultants;
- Engineers, architects, and other design professionals;
- Educators; and
- Researchers.

Although there are wide range of objective complexities covered in the chapters, a deep knowledge of these objective topics is not required to achieve familiarity and benefit from the content. For readers who may not have the time to go in depth in each subject matter, it is suggested that they initially become familiar with the "recommended practices" at the end of each chapter. Each reader can then look at the chapter in depth to learn the reasonings/sources of these recommended practices.

Note that ASCE Manuals of Practice (MOPs) are developed by ASCE technical committees, such as the ORC, under the direction of an ASCE sponsor such as the Engineering Mechanics Institute (EMI). The distinguishing characteristic of an MOP, including this one, is that each one undergoes peer review by a Blue Ribbon Panel of experts before final approval is sought from the appropriate executive committee. Thus, the peer review by the Blue Ribbon Panel gives added weight to the MOP.

Mohammed M. Ettouney, Ph.D., P.E., F.AEI, Dist.M.ASCE

REFERENCES

NIAC (National Infrastructure Advisory Council). 2009. *Critical infrastructure resilience final report and recommendations.* Washington, DC: NIAC.

NSC (National Security Council). 2011. "Presidential Policy Directive/PPD-8: National Preparedness." *Presedential Policy Directive.* Accessed May 26, 2018. https://www.dhs.gov/presidential-policy-directive-8-national-preparedness.

Office of the Press Secretary. 2013. "Presidential Policy Directive/PPD-21: Critical infrastructure security and resilience." *Presedential Policy Directive.* Accessed October 20, 2019. https://www.dhs.gov/sites/default/files/publications/PPD-21-Critical-Infrastructure-and-Resilience-508.pdf.

CHAPTER 1

ON THE DEFINITION OF RESILIENCE

S. Gerasimidis, M. Ettouney

1.1 INTRODUCTION

Natural and human-made hazards in recent years have led to disasters causing significant damage to communities and their infrastructure. The associated losses with these disasters have illustrated the need for designing civil infrastructure that can withstand such events with minimum disruption. The approach of investigating civil infrastructure against disasters in terms of losses and consequences, continuation of operations, and time to recovery (The terms "time to recovery," "rapid recovery," or just "recovery" have been used by different authors. We will use them interchangeably for the remainder of this chapter.) has been widely known as "infrastructure resilience." However, because several researchers and agencies have approached infrastructure resilience from various viewpoints, it is considered beneficial to provide a closer look at the definition of the term "resilience." This is the main objective of this chapter and one of the goals of the current Manual of Practice.

This chapter is not intended to be a complete inventory of all the definitions of resilience that have appeared in the literature. Not only would this be a pointless long inventory of definitions, it is the authors' opinion that it would lead to an outcome opposite of the intended one. The intention of the authors is to highlight some important definitions and identify common patterns overarching almost all definitions. Along these lines, this chapter does not aim at adding one more definition to the ones that already exist in the literature. The approach adopted herein is to provide useful key properties and common components among all measurements and definitions of resilience that can describe the concept of resilience for infrastructure systems, distinguishing resilience from

1

other related but not identical concepts such as risk, reliability, and sustainability. The chapter also investigates the attributes of resilience definitions, especially from the viewpoints of objective processes. To help in this investigation, we also introduce a unifying resilience definition theory [theory of resilience definitions (TRD)].

The chapter is organized as follows. In Section 1.2, several key observations on the definition of resilience are presented to form a background for the concept of resilience. These key observations aim to provide the motivation and the need for a study of the definition of resilience. In Section 1.3, a historical perspective of the use of the term resilience is presented from the early uses in engineering manuscripts to the extensive use of the term today. Section 1.4 presents a thorough study on most of the main key definitions that appear in the literature with an emphasis on infrastructure resilience. This section also includes definitions published in reports from different agencies. Section 1.5 aims at identifying and presenting themes among the different definitions of resilience provided in Section 1.4 and found in the literature. The important relationship between the concepts of risk and resilience is described in Section 1.6. We discuss the attributes of resilience definitions in Section 1.7 and introduce TRDs that might prove to be of help to analysts and decision makers who are embarking on an objective resilience effort. The chapter's conclusions are presented in Section 1.8 and best/recommended practices are given in Section 1.9.

1.2 KEY OBSERVATIONS FOR THE DEFINITION OF RESILIENCE

The concept of infrastructure resilience and resilience, in general, has found a wide range of applications in many disciplines. As it will be shown in the next section, although the term resilience initiated from engineering disciplines, the beginning of the era of extensive use started from disciplines quite distant to engineering. Recently, infrastructure resilience has gained significant momentum mainly because of the immense losses associated with natural and man-made disasters. In this environment, there are a plethora of entities that are approaching resilience and its assessment in an effort to provide an adequate definition. This section provides answers and observations to key questions regarding the definition of resilience, such as the following: Why do we need a definition for resilience? Is resilience a useful concept? Is resilience a desirable attribute? Does enhanced resilience mean lower vulnerability, lower risk, or not?

It is considered important to present a comprehensive overview of the definitions in the literature and identify commonalities and key characteristics. If we aim to be objective about the assessment and management of resilience, the first step is to identify and define the area of interest. Without a comprehensive definition, it becomes almost impossible to proceed to formalization and development of operational tools. It has already been observed by researchers in the field that the main problem with resilience is the multitude of different definitions and turning any of them into operational tools (Klein et al. 2003). The same group of authors go one step further to state that after many years of debate and analysis, the definition of resilience has become so broad as to render it almost meaningless. This has led to considerable confusion among interested communities, resulting in a difficulty to produce practical policy or management tools. Therefore, there is a significant need for a comprehensive definition of resilience and in particular of infrastructure resilience.

The usefulness of the concept of resilience can be identified through the different components of resilience that have already been proposed in the literature.

1.2.1 Assessment of Resilience

The two most important elements within the assessment of resilience can be defined as operational state and time to recovery. These elements describe the way an infrastructure system would respond to an extreme event and whether it regains the desired operational state in a reasonable time. These ideas entail concepts of losses and processes of how the system will come back to an acceptable operational state and are directly connected to the commonly known 4Rs: robustness, resourcefulness, recovery, and redundancy.

1.2.2 Acceptance of the Assessment

However, resilience assessment is only the first necessary component of the comprehensive definition of infrastructure resilience. Even if we develop resilience assessment tools, the impact of these tools becomes very limited if they do not belong to a comprehensive resilience framework that provides essential context for the assessment. As pointed out in Ettouney (2014), only a resilience management approach can provide a comprehensive definition of resilience that incorporates how different stakeholders can manage the resilience of their assets or communities. Following resilience assessment, the second major component is the acceptance of this assessment. If the assessment of resilience indicates that the level of resilience is acceptable, then the system is considered adequately resilient.

1.2.3 Resilience Improvement

In case the level of resilience is not acceptable, the third component of resilience management, resilience improvement, comes into the picture,

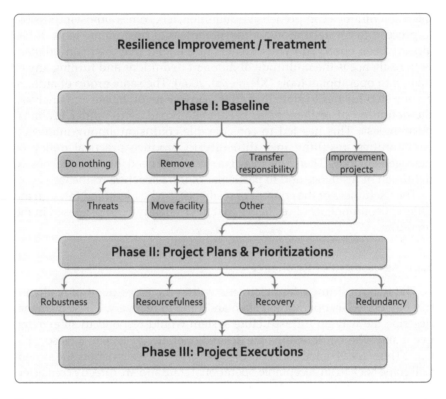

Figure 1-1. An example of the different characteristics of resilience improvement. Source: Adapted from Ettouney (2014).

which refers to processes and investments needed to bring the level of resilience to an acceptable level (Figure 1-1).

1.2.4 Resilience Monitoring

The fourth resilience management component is known as "resilience monitoring," which refers to the importance of knowing the level of resilience of the asset in time. Time tends to reduce the physical capacity of infrastructures as well as their operational level and resilience, whereas demands tend to increase them.

1.2.5 Communication

Finally, the fifth and last component of resilience management is the communication of all the previously mentioned activities to the stakeholders, the legislators, and eventually the public, which would aid in securing awareness, confidence, and the necessary resources for

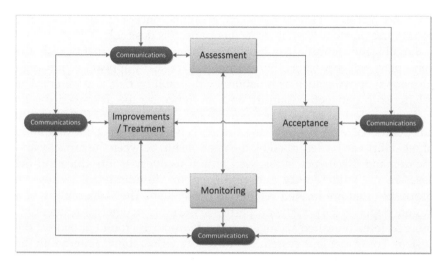

Figure 1-2. An overview of the resilience management components and their interconnectivities.
Source: Adapted from Ettouney (2014).

achieving resilience. Therefore, there is significant value in the effort to provide a definition for the concept of resilience, especially after moving from the realm of assessment into management, accounting for all the interconnectivities among components (Figure 1-2). Through resilience management, resilience can eventually lead to policy making and useful operational tools.

Another key observation comes from whether resilience is a desirable attribute of systems. Although instinctively resilience is considered to be always desirable, an interesting approach to that question has been presented in Klein et al. (2003), stating that whether resilience is a desirable attribute depends also on the definition of the concept. In the same paper, the authors use the example of a city-urban system that is struck by a disaster. They mention that if a city struck by a disaster is experiencing extensive losses and consequences, then going back to the original predisaster state is undesirable because it would leave the city as vulnerable to the disaster as in the first place. Therefore, the common definition of resilience as the property of a system to bounce back is very limiting and does not tell the whole story. However, if the definition of resilience includes the acceptance level as previously proposed through the lens of resilience management, resilience becomes a desirable attribute of the system. An additional beneficial component of resilience, in that case, would also be the ability of a system to build and increase the capacity for learning and adapting to new demands, which is connected to the monitoring component of management. Therefore, the definition of resilience has such a fundamental effect on the concept of resilience that it could force one to ask whether resilience itself is desirable.

Finally, the definition of resilience is important to distinguish the concept of resilience from other attributes that usually appear as equivalent ideas or synonyms such as reliability, vulnerability, risk, and others. The relationship between reliability and resilience can be described by the main characteristic of resilience referring to the recovery of the system to an operational level. For infrastructure and civil engineering, the definition of reliability can be described by the relationship between the capacity and the demand. Resilience is assessed only if recovery is introduced in the assessment. Vulnerability and resilience are differentiated because of properties that are located relative to the system. The vulnerability of a system depends only on the properties of the system, whereas resilience is more an interconnected arrangement with properties found outside of the system. For example, a system can have low vulnerability (internal to the system), but its failure can have huge consequences (external to the system). The most interesting connection of all these is probably the one between risk and resilience. This chapter includes a separate section closer to the end on this topic and how risk is a superset of resilience.

1.3 THE TERM "RESILIENCE" AND A HISTORICAL PERSPECTIVE OF THE APPEARANCE OF THE GENERAL CONCEPT OF RESILIENCE - ETYMOLOGY

As a starting point, the etymology of the word resilience is acquired from the field of linguistics in an effort to identify where the word comes from. A general consensus on the origin of the word suggests that the word originates from the Latin word *salio*, which means to jump or attempt to jump back. The addition of the preposition "re" to *salio* gives *resilio*, and in Latin this would mean to make something emerge or to make something come back (personal communication, Babiniotis, 2015). Based on all this, a first short conclusion would be that the linguistic definition of the word resilience is the reappearance, the renaissance, or the reinstatement aiming at the restoration of a previous better state or condition. In other words, the definition can be the reactivation of something so that it can be used again. A widely used English language dictionary, Merriam-Webster (2016), states that resilience is the capability of a strained body to recover its size and shape after deformation caused especially by compressive stress, or an ability to recover from or adjust easily to misfortune or change.

The frequency of the use of the word resilience in the literature is shown in Figure 1-3 through an Ngram extracted by Google (2020). It can be seen that there has been a dramatic increase in the usage of the term resilience in last few decades. Interestingly, there is an increase in its usage in the

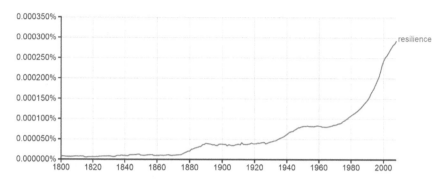

Figure 1-3. Ngram from google showing the use of the word resilience in time.

decades between 1870 and 1890 that originates from the field of engineering mechanics. During that era, there were several references that used the term resilience as another expression for the "amount of work done in breaking an alloy," such as those mentioned from US Board to Test Iron, Steel and Other Metals (1879). Love (1893) includes the following reference to the word: "The term resilience was introduced into Physics by Thomas Young with the definition—'The action which resists pressure is called strength, and that which resists impulse may properly be termed resilience.' The resilience of a body is jointly proportional to its strength and its toughness and is measured by the product of the mass and the square of the velocity of a body capable of breaking it, or of the mass and the height from which it must fall in order to acquire this velocity; while the strength is merely measured by the greatest pressure which it can support." The mentioning of the word resilience in the book by Young (1807) is one of the earliest found by the authors of this chapter.

Russell (1898) uses the term resilience as a limit state: "When stress is applied to a solid body, the material is distorted and a certain amount of work or energy is absorbed. The work thus absorbed in the deformation of the material is called resilience. If the stress changes from zero up to the elastic limit of the material, the energy absorbed during the change is the 'elastic resilience' of the material. If the stress changes from zero up to the ultimate strength of the body, the energy absorbed is the 'ultimate resilience' of the body." In a note from Russell (1898), it is stated: "This use of the word resilience will be objected to by some as not being in conformity with the original meaning of the word. It is sanctioned, however, by some authorities (see Thurston's 'Materials of Engineering') and for want of a good substitute, may be considered as a technical term."

The second significant increase in the use of the word appears to be in the decades 1930 to 1950. In that era, the term was used with the same meaning as in previous years, describing the work done in loading a specimen. Finally, during the decade of 1980, the use of the term literally took off until today. It is then that alternative uses of the word appear mainly in human psychology and in ecology and until today when resilience has entered the field of infrastructure and community resilience to hazard events.

A typical example from the field of ecological resilience is the definition given by Walker and Meyers (2004): "The capacity of an ecosystem to respond to a perturbation or disturbance by resisting damage and recovering quickly." In other words, resilience is defined as the capacity of a system to absorb disturbance and reorganize while undergoing change so as to retain essentially the same function, structure, identity, and feedbacks. Another example is given by Folke et al. (2002), who also used resilience for social–ecological systems (SESs). They provide a definition relating resilience to the following: "The magnitude of the shock that the system can absorb and remain within a given state, the degree to which the system is capable of self-organization and to the degree to which the system can build capacity for learning and adaptation." According to Berkes et al. (2003), the idea was to focus not merely on ecosystems per se or societies per se but on the integrated social–ecological system.

The second field that has used the term resilience extensively is the field of psychology in the sense of psychological resilience. In this case, the American Psychological Association (APA 2014) provides the following definition: "The process of adapting well in the face of adversity, trauma, tragedy, threats or significant sources of stress — such as family and relationship problems, serious health problems or workplace and financial stressors. It means 'bouncing back' from difficult experiences. Being resilient does not mean that a person doesn't experience difficulty or distress. Emotional pain and sadness are common in people who have suffered major adversity or trauma in their lives." The same reference continues to suggest that, in fact, the road to resilience is likely to involve considerable emotional distress: "Resilience is not a trait that people either have or do not have. It involves behaviors, thoughts and actions that can be learned and developed in anyone."

Finally, the word resilience has entered the field of infrastructure resilience with a typical example broad definition given by Zhou et al. (2010): "The capacity to resist and recover from loss, is an essential concept in natural hazards research and is central to the development of disaster reduction at the local, national and international levels."

1.4 GENERAL KEY DEFINITIONS OF RESILIENCE

This section includes a brief reporting of definitions of resilience that come from different disciplines but tend to be independent of their origin. The goal of this section is to identify common patterns and key properties that will be later described in Section 1.5. The first part of this section includes definitions from the literature of journal papers, and the second part includes the definitions provided by National Agencies. It is noted here that this reporting is not exhaustive, and the authors acknowledge that there are numerous references with other definitions of resilience that are not reported here. As stated previously, the goal is not to provide a complete list of definitions, a task that would be almost impossible to complete, but an overview of how the term resilience has evolved in time by highlighting representative definitions in time and capturing the basic components that were common to each.

1.4.1 Definitions from the Literature

The first general definition of resilience is provided by Holling (1973), who defines resilience as the amount of disturbance that can be sustained by a system before a change in system control or structure occurs. It could be measured by the *magnitude of disturbance* that the system can tolerate and still persist. Gordon (1978) defines resilience using a purely mechanics approach as the ability to store energy and deflect elastically under a specified loading condition without breakage or deformation. A little later, Timmerman (1981) suggests that resilience is the ability of human communities to *withstand external shocks* or perturbations to their infrastructure and to *recover* from such perturbations. Pimm (1984) highlighted the relationship of resilience to recovery by observing that resilience is the speed with which a system *returns to its original state* following a perturbation. The concept of bouncing back is demonstrated by Wildavsky (1991), who says that resilience is the capacity to *cope with unanticipated dangers* after they have become manifest, learning to *bounce back*. Similarly, Holling et al (1995), in an updated definition, suggest that resilience is the buffer capacity or the ability of a system to *absorb perturbation*, or the magnitude of disturbance that can be absorbed before a system changes its structure by changing the variables. Horne and Orr (1998), talking about organizations in general, offered a broader definition: Resilience is a fundamental quality of individuals, groups, and organizations and systems as a whole to *respond productively* to significant change that disrupts the expected pattern of events without engaging in an extended period of regressive behavior. Regarding disasters, Mileti (1999) states that local resiliency with regard to disasters means that a locale is able to *withstand an extreme natural event* without suffering devastating losses, damage, diminished productivity, or quality of life without a large amount of assistance from outside the community. In the context of disaster management, resilience is used to describe the *ability to resist or adapt to stress* from hazards and the ability to recover quickly.

Coming from the social and ecological resilience perspective, Adger (2000) decomposes the concept of resilience and makes the observation that resilience is understood as having three properties: resistance, recovery, and creativity, in which (1) *resistance* relates to a social entity's efforts to withstand a disturbance and its consequences and can be understood in terms of the degree of disruption that can be accommodated without social entity undergoing long-term change. (2) *Recovery* relates to an entity's ability to pull through the disturbance and can be understood in terms of the time taken for an entity to recover from a disruption. (3) *Creativity* is represented by a gain in resilience achieved as part of the recovery process, and it can be attained by adapting to new circumstances and learning from the disturbance experience. Originating from the same field, Carpenter et al. (2001) state that the resilience alliance (social–ecological organization) consistently refers to SESs and defines their resilience by considering three distinct dimensions: (1) The amount of disturbance a system can absorb and still remain within the same state or domain of attraction; (2) the degree to which the system is capable of self-organization; (3) the degree to which the system can build and increase the capacity for learning and adaptation. Focusing on the recovery part of resilience, Paton and Johnston (2001) define resilience as the ability to pick up and utilize physical and economic resources for effective recovery following hazards.

A pivotal point in the thread of resilience definitions is the extensive definition provided by Bruneau et al. (2003). First, this definition identifies resilience at four levels: (1) *technical*, how physical systems perform; (2) *organizational*, the ability to respond to emergencies and carry out critical functions; (3) *social*, the capacity to reduce the negative social consequences of loss of critical services; and (4) *economic*, the capacity to reduce both direct and indirect economic losses. Second, four dimensions of resilience are provided: (1) *robustness*, the strength to withstand a given level of stress without loss of function; (2) *redundancy*, the extent to which elements and systems are substitutable; (3) *resourcefulness*, the capacity to identify problems, establish priorities, and mobilize resources; and (4) *rapidity*, the capacity to meet priorities and achieve goals in a timely manner. By this definition, a resilient system has reduced probability of failures, reduced consequences from failures, and reduced time to recovery.

A little later, Walter (2004) provides another definition connecting losses, natural disasters, and networks: "Resilience is the capacity to survive, adapt, and recover from a natural disaster. Resilience relies on understanding the nature of possible natural disasters and taking steps to reduce risk before an event as well as providing for quick recovery when a natural disaster occurs. These activities necessitate institutionalized planning and response networks to minimize diminished productivity, devastating losses, and decreased quality of life in the event of a disaster."

Studying water service disruptions, Rose and Liao (2005) state that resilience is the adaptive response to hazards to enable individuals and communities to avoid potential losses.

Regarding social resilience, Maguire and Hagan (2007) state that social resilience is the capacity of a social entity, for example, group or community, to bounce back or respond positively to adversity. Similar to definitions coming from other aspects of resilience, they identify that social resilience has three major properties: resistance, recovery, and creativity. Cutter et al. (2008) define resilience as the ability of a social system to respond and recover from disasters and include those inherent conditions that allow the system to absorb impacts and cope with an event, post event, and adaptive processes that facilitate the ability of the social system to reorganize, change, and learn in response to a threat.

Figure 1-4 illustrates graphically the concept of resilience through a comparison between two systems, emphasizing on the robustness and recovery aspects.

1.4.2 Definitions from Agencies

The Transportation Research Board, along with the US National Academies (NRC 2012), has defined resilience as the ability to prepare and plan for, absorb, recover from, or more successfully adapt to actual or potential adverse events. Following a very similar approach, the Department of Homeland Security (DHS 2009) has provided a definition for resilience as the ability to resist, absorb, recover from, or successfully adapt to adversity or a change in conditions and a very similar one a little later DHS (2015) as the ability to adapt to changing conditions and withstand and rapidly recover from disruption owing to emergencies. Another definition comes from the United Nations/International Strategy for Disaster Reduction (UN/ISDR 2002), which states that the capacity of a system, community, or society potentially exposed to hazards to adapt by resisting or changing to reach and maintain an acceptable level of functioning and structure. This is determined by the degree to which the social system is *capable of organizing itself* to increase this capacity for learning from past disasters for better future protection and to improve risk reduction measures (NRC 2012).

The PPD-8 (Presidential Policy Directive), see National Security Council (NSC 2011), defined resilience as "the ability to adapt to changing conditions and withstand and rapidly recover from disruption due to emergencies." Later, the PPD-21, see Office of the Press Secretary (OPS 2013), extended the definition slightly to "the ability to prepare for and adapt to changing conditions and to withstand and recover rapidly from disruption. Resilience includes the ability to withstand and recover from deliberate attacks, accidents, or naturally occurring threats or incidents."

We note that a popular definition of "disaster" was issued previously by the National Science and Technology Council (NSTC 2005) as "a serious disruption of the functioning of a community or society causing widespread human, material, economic or environmental losses which exceed the ability of the affected community or society to cope using its own resources." NIST (2015, 2016) used these definitions in their community resilience planning guides.

The National Infrastructure Advisory Council's Critical Infrastructure Resilience (NIAC 2009) provides an extensive definition of resilience: Infrastructure resilience is the ability to reduce the magnitude and/or duration of disruptive events. The effectiveness of a resilient infrastructure or enterprise depends on its ability to anticipate, absorb, adapt to, and/or rapidly recover from a potentially disruptive event (NIAC 2009), as measured through the following four attributes: *Robustness*: the ability to maintain critical operations and functions in the face of crisis. This includes the building itself, the design of the infrastructure (office buildings, power generation, distribution structures, bridges, dams, levees), or system redundancy and substitution (transportation, power grid, communications networks) (NIAC 2009). *Resourcefulness*: the ability to skillfully prepare for, respond to, and manage a crisis or disruption as it unfolds. This includes identifying courses of action and business continuity planning, training, supply chain management, prioritizing actions to control and mitigate damage, and effectively communicating decisions (NIAC 2009). *Rapid recovery*: the ability to return to and/or reconstitute normal operations as quickly and efficiently as possible after a disruption. Components of rapid recovery include carefully drafted contingency plans, competent emergency operations, and the means to get the right people and resources to the right places (NIAC 2009). *Redundancy*: backup resources to support the originals in case of failure (Ettouney 2014). These four resilience components: robustness, resourcefulness, rapid recovery, and redundancy, which are simply referred to as the 4Rs, have been used by many authors, especially while addressing resilience assessments in an objective manner (Kennett et al. 2011a, b, c).

Although not an agency, a coalition of building industries issued a joint statement in 2014 (updated in 2016) on resilience and how it impacted the building industry. The coalition included over 38 members of the building industry. The Resilience Building Coalition also released a set of guiding principles to help the building industry adopt resilient design and policies. These include developing and advocating for codes and policies that advance resilience; developing "whole-systems resilient design" approaches for the built environment; and providing guidance, beyond the baseline life-safety codes, that recognizes the importance of fortifying property for individual and community resilience (Industry Statement on Resilience 2016).

1.5 KEY PROPERTIES AND COMMON COMPONENTS OF RESILIENCE: UNIVERSAL RESILIENCE DEFINITION

All the aforementioned definitions aim at describing the key attributes of the concept of resilience, and although many of them approach resilience from different viewpoints, there are several common themes and characteristics that can be identified and are summarized here.

Among the most important properties of resilient systems are the ability to continue operations and the ability to recover to an accepted level of operations in a timely manner. The desirable targeted operational level after recovery could be the original state or an adjusted state and this depends on the resilience acceptance. An additional desirable property of a resilient system would be to minimize the time to return to a global equilibrium even if this would not mean the original state but intermediate equilibrium states that would allow organization and minimizing the losses of the system in terms of damage or operational capacity. This would also mean that the system should be able to withstand the direct effect of the damaging event without losing the capacity to allocate resources and initiate a recovery process.

Equally important to the continuation of operations and recovery is the property to sustain a shock and have the capacity to survive it. Without survival and the opportunity to recover, resilience has no meaning. This would mean to minimize the chances to have a complete failure of the system to an acceptable level and to minimize the amount of disturbance before any change of state.

As stated previously, all reviewed resilience definitions incorporated these two all-important attributes: reduced losses of operational capabilities (the authors use different descriptors for this subcomponent such as quality of operations, functionality, and so on) and rapid recovery. Because of this, we consider a universal resilience definition as a top-level definition that includes these two attributes.

In addition to the universal resilience definition, most definitions venture into a second tier of attributes that aims at identifying subcomponents of the two top-level attributes. It is in attempting to define these subcomponents where different resilience definitions start to diverge. For example, a desirable attribute is to maintain a minimum level of service while undergoing changes and the capability to maintain an acceptable functional level during recovery. Extending this idea would involve retaining relationships between people or state variables and improving the reorganizational capacity of the system.

Additional second-tier resilience components require that a resilient system would be capable of anticipating, adapting, and recognizing unanticipated perturbations and also adjusting to unexpected threats. Through coordinated planning, a resilient system should be able to prepare for future protection efforts and continuously learn and adapt to new demands. Structural health monitoring is a key aspect in the ability to become aware of the degradation of the system through time and the level of resilience as a function of this degradation. We note that all of those second-tier components can be traced into one, or both, of the two top-level components: minimal loss of operational capacity and/or rapid recovery.

Before we end our discussion of the properties of resilience definitions, we need to discuss the subjects of system definition and hazards. We note that some resilience definitions include a definition of the system they are targeting. Some other definitions are more general and are meant for generic applications. The system can be a specific single asset, such as a building, a bridge, or a tunnel. Other systems can be much larger, such as a network of assets, such as a health campus, transportation network, or school campus. Even larger systems might be considered, such as a community of assets, such as a city, state, or country. Similarly, resilience definitions might target specific hazards, such as earthquakes, floods, or terrorism. Some other definitions simply mention hazards in general terms such as "natural" or "man-made" hazards. Even a more general mention of "disasters" has been used in some definitions. The magnitude/severity of the hazard is usually left out of definitions. The level of the hazard is usually left to decision makers to determine. We note that even though target systems and types of hazards may or may not be included in the definition of resilience, they play an essential role during any resilience management effort, including objective resilience management processes. As such, they need to be clearly defined during such efforts or processes.

Figure 1-5 illustrates the concept of top-tier and second-tier resilience definitions as well as their relationship with the system and hazard.

1.6 RESILIENCE VERSUS RISK

The relationship between risk and resilience is particularly important as it possibly paves the way for objectifying resilience. Risk has been a concept studied for a long time, certainly much more than resilience, by many communities (outside engineering too, financial, insurance, security, economics). Over the years, all these communities have accumulated an immense wealth of objective and subjective methodologies for computing risk. By looking at the relationship between resilience and risk and essentially by defining resilience as a subset of risk, we can easily use advanced risk methodologies and processes with a little bit of modification and apply them to objectify resilience.

A common risk definition can be found in Ettouney (2016) and Ettouney and Alampalli (2016): Risk is a description of an uncertain alpha-numeric

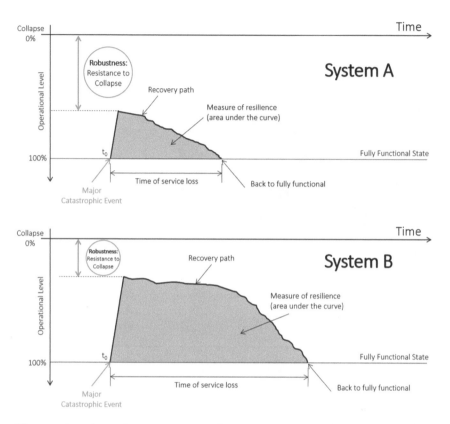

Figure 1-4. Schematic representation of resilience. System A is more resilient than System B.

expression (objective or subjective) that describes an outcome of an unfavorable uncertain event that might degrade the performance of a single (or of a community) civil infrastructure asset (or assets). In other words, risk is the relationship between a particular hazard (or threat) that might degrade the performance of the infrastructure under consideration and the consequences that might result from degradation of performance (Gutteling and Wiegman 1996; FEMA 2005; NRC 2010). Essentially, the main characteristic of risk is the cost related to the consequences of the degradation.

On the contrary, two main themes in the definition of resilience are the time to recover and, in general, continuation of operations (including, but not limited to, time to recover). Therefore, objective resilience is also described in terms of cost (monetary and other expressions of it). However, the cost associated with resilience regards purely the continuation of operations and time to recovery to an acceptable state that can be considered a subset of the more general risk consequences that are not

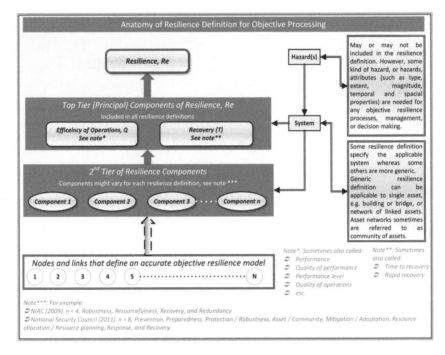

Figure 1-5. Components of resilience definitions.

limited to only continuity of operations and time to recovery. For example, costs of personal material losses as a result of a storm can be considered as risk consequences, but they, or most of them, are not a function of continuity of operations or time to recover.

That is why resilience should be considered to be a subset of risk and not the opposite and the relationship is inverse where increased resilience decreases risk but decreased resilience increases risk. Schematically, the relationships among risk, resilience, and other concepts such as reliability or sustainability are shown in Figure 1-6. In this figure, it is shown that risk has three main components: vulnerability, hazard, and consequences that are interconnected to the 4Rs of resilience.

1.7 NEEDED ATTRIBUTES FOR OBJECTIVITY AND THEORY OF RESILIENCE DEFINITION

1.7.1 Needed Attributes for Objectivity

One of the reasons we are interested in studying resilience definitions is that we recognize the essentiality of definitions to any objective treatment

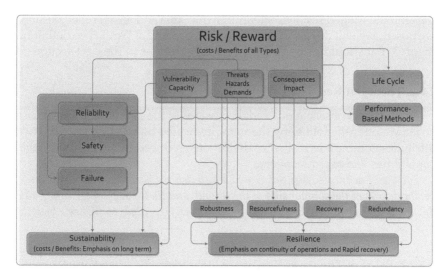

Figure 1-6. The relationships among risk, resilience, reliability, and sustainability. Source: Adapted from Ettouney (2014).

of resilience. Without an adequate definition of resilience, we cannot hope to objectively formalize resilience accurately or consistently. This brings us to an apparent dilemma: given the plethora of resilience definitions we have just reviewed and the wide divergence of the descriptions of the resilience components of those definitions, does this mean that objective resilience will accordingly produce differing results that are dependent on the particular definition that was used to define the objective process? If true, then another question might arise: Which resilience definition will produce a more accurate, objective result?

To address these two questions, and the apparent dilemma they raise, we need to ask ourselves: What are the attributes for an accurate, objective resilience process? It turns out that the answer to such a question is fairly simple. The attributes of an accurate objective resilience process or any other objective process at large are as follows: (1) Comprehensive treatment. The process needs to include, as its independent variables, all important governing issues that might affect the needed outcome or outcomes, such as an estimation of the resilience metric itself, *Re*. If the objective process does not include one of those underlying variables, then the resultant outcome will not be accurate. (2) Integral considerations. In addition to including all important variables, the links (sometimes called interconnectivities) among those variables need to be objectively well handled in the objective process.

Armed with these two essential attributes (comprehensiveness and integrality) for objective resilience processes, we are now ready to address the aforementioned resilience definition dilemma and answer the questions we posed. We will do so by proposing, then proving, a "resilience definition theory" that should provide the promised solutions in the next section.

1.7.2 Theory of Resilience Definition

Based on our observations and discussions of the definitions of resilience so far, we offer the following theorem predicated on acceptance of the two attributes common in resilience definitions: reduced losses of operational capabilities and rapid recovery.

For a well-defined system set, S, and a well-defined hazard, or a set of hazards, H, the details of the components of the second, and their subcomponents, if any, do not alter significantly the results of a well-executed resilience objective process.

1.7.2.1 Proof of Theory of Resilience Definition. *Assumptions:* We make the following assumptions for the TRD to be valid: (1) We first assume that for any desired resilience-related objective process, S and H are modeled adequately by the set $G_0 = G_0(V,E)$, with V being a set of variables that describe S and H in a comprehensive manner (with a resolution that is pertinent to the objective process and its goals). The set E represents adequate links (interconnections) between the population of V so as to satisfy the integral requirement of Section 7.1. This means that G_0 is a satisfactory model that should be expected to produce a well-executed objective resilience result. (2) As a direct result of Assumption 1, it is logical to expect that G_0 includes enough representation of variables controlling the two first-tier components: quality of operations, Q, and rapid recovery, T. Note that many members of G_0 can control both Q and T. Further, we introduce the following two sets:

$$G_{11} = G_{11}(Q) \tag{1-1}$$

$$G_{12} = G_{12}(T) \tag{1-2}$$

such that

$$G_0 = G_{11} \cup G_{12} \tag{1-3}$$

Again, the model of the aforementioned equation is assumed to produce an adequate objective resilience result.

Need for second-level resilience definition components: Let us assume that for a particular resilience definition, $\mathcal{A}$ contains an N_A second tier of components, $G_{A2}|_i$, with $i \in N_A$.

Without any loss of generality, we can state that

$$G_0 = G_{A2}\big|_1 \cup G_{A2}\big|_2 \cup G_{A2}\big|_3 \cup \cdots G_{A2}\big|_{N_A} \tag{1-4}$$

Similarly, for another resilience definition, $\mathcal{B}$ contains an N_B second tier of components, $G_{B2}\big|_i$, with $i \in N_B$.

Again, without any loss of generality, we can state that

$$G_0 = G_{B2}\big|_1 \cup G_{B2}\big|_2 \cup G_{B2}\big|_3 \cup \cdots G_{B2}\big|_{N_B} \tag{1-5}$$

Let us assume further that both $\mathcal{A}$ and $\mathcal{B}$ contain an adequate representation of the two basic resilience components Q and T, and that both sets $G_{A2}\big|_i$ and $G_{B2}\big|_i$ have enough information and components to produce a well-executed objective resilience result. Then, from Equations (1-4) and (1-5), both $\mathcal{A}$ and $\mathcal{B}$ are capable of producing similar objective resilience results. The same logic can be used to prove the TRD if a third or fourth subcomponent definition tiers are introduced in the resilience definitions, as long as the components of the tier that is used for the objective process produce well-executed objective resilience results.

1.7.2.2 Implications of Theory of Resilience Definition. The TRD has some important implications for any objective resilience processes, and these are as follows: (1) If the resilience objective model under consideration is comprehensive and integral, see Section 1.7.1, and if that the model satisfies the requirements of the first resilience definition tier, then it should produce adequate results, no matter what the subjective resilience components of the second or higher tiers state. (2) The aforementioned indicates that if the resilience definition on hand allows for the first-tier components, then it should be equivalent to any/all other resilience definitions that have similar allowance, from an objective processing viewpoint.

We are in a position now to ask about the value if any of higher resilience definition tiers beyond the first tier, because the TRD states that they are not needed to produce adequate objective results. In practice, such subcomponent definitions of higher tiers are valuable during any objective resilience process. These subcomponents can act as a guide while forming the objective model G_0 and its subcomponents V and E. They can also be of great help in forming the components of S and H. Our purpose of introducing the TRD is to remove any confusion about the utilization of the plethora of resilience definitions: They are all sufficient to produce an adequate objective process. However, those definitions with more tiers might be more helpful in the modeling process. As such, the modelers and decision makers might choose the definition that can be more helpful in achieving their ultimate objective tool, with the knowledge, according to the TRD, that they are not missing any value by not following other definitions.

1.8 SUMMARY AND CONCLUSIONS

The popularity of the resilience paradigm in recent years promoted proliferation of definitions of the term resilience. This led, in many cases, to some confusing usage of the term and conflating it with other equally popular paradigms in the field of civil engineering, such as safety, failure, reliability, risk, and/or sustainability. This chapter aimed at tracing the definitions of resilience over the years up to more recent definitions.

We show that even though most of the resilience definitions utilize different wordings, they are mostly, in a generic sense, similar in their intentions and implications. Resilience definitions, especially recent ones, specify different components. However, on a closer examination, we can find that these components actually do intersect with each other to such an extent that if the goal of the users is the objective treatment of resilience, the ultimate input variables that identify resilience, or any of its basic components, should be largely similar to such an extent that the differences among the underlying resilience definitions are not a controlling factor in the ultimate sought-after objective result. Of course, those resilience components, and their underlying input variables, need to be comprehensive and integral. These two requirements should be expected, though, because if either of them is lacking in the objective process, the ultimate result would not/should not be expected to be accurate.

1.9 RECOMMENDED PRACTICES

Based on our findings in this chapter, a few recommended practices and future research needs emerge as follows:

- Concept of resilience is useful as it can lead to optimal policies and strategies.
- There is a need for the community to extend the concept of resilience from a purely assessment point of view to a resilience management point of view.
- Most recent resilience definitions are similar in their underlying aims and implications, even though they vary in their text wordings. Following comprehensive and integral objective processes can lead to similar objective resilience management results, as proven by the TRD.

These include assessment, acceptance, improvement, monitoring, and communication results.

- Safety, failure, and reliability (that relate capacity and demands) are not synonymous with resilience. These concepts are only a subset of resilience. Embarking on a resilience-based effort should account for all resilience components, in addition to limiting concepts.
- Comprehensive and integral resilience-based efforts are a multidisciplinary effort. There is an essential need for coordination among different disciplines, otherwise, a less than optimal performance might result.
- Linking resilience components (of any of the recent popular definitions) and different objective processes are required.

REFERENCES

Adger, W. N. 2000. "Social and ecological resilience: Are they related?" *Prog. Hum. Geogr.* 24 (3): 347–364.

APA (American Psychological Association). 2014. *The road to resilience.* Washington, DC: APA.

Berkes, F., J. Colding, and C. Folke, eds. 2003. *Navigating social–ecological systems: Building resilience for complexity and change.* Cambridge: Cambridge University Press.

Bruneau, M., S. Chang, R. Eguchi, G. Lee, T. O'Rourke, A. Reinhorn, et al. 2003. "A framework to quantitatively assess and enhance the seismic resilience of communities." *Earthquake Spectra* 19 (4): 733–752.

Carpenter, S., B. Walker, J. M. Anderies, and N. Abel. 2001. "From metaphor to measurement: Resilience of what to what?" *Ecosystems* 4 (8): 765–781.

Cutter, S. L., L. Barnes, M. Berry, C. Burton, E. Evans, E. Tate, et al. 2008. "A place-based model for understanding community resilience to natural disasters." *Global Environ. Change* 18 (4): 598–606.

DHS (Department of Homeland Security). 2009. *National infrastructure protection plan—Partnering to enhance protection and resilience.* Washington, DC: DHS.

DHS. 2015. *National infrastructure preparedness goal.* Washington, DC: DHS.

Ettouney, M. 2014. *Resilience management: How it is becoming essential to civil infrastructure recovery.* New York: McGraw Hill.

Ettouney, M. 2016. "Resilience and risk management." Accessed March 2021. http://c.ymcdn.com/sites/www.nibs.org/resource/resmgr/Conference2016/BI2016_0113_ila_ettouney.pdf.

Ettouney, M., and S. Alampalli. 2016. *Risk management in civil infrastructure.* New York: CRC Press.

FEMA. 2005. *Risk assessment: A how-to guide to mitigate terrorist attacks, risk management series.* FEMA 452. Washington, DC: FEMA.

Folke, C., S. Carpenter, T. Elmqvist, L. Gunderson, C. S. Holling, and B. Walker. 2002. "Resilience and sustainable development: Building adaptive capacity in a world of transformations." *AMBIO* 31 (5): 437–440.

Google. 2020. "Google Ngram Viewer." https://books.google.com/ngrams.

Gordon, J. 1978. *Structures or why things don't fall down.* Harmondsworth, UK: Da Capo Press/Penguin Books.

Gutteling, J., and O. Wiegman. 1996. *Exploring risk communications.* Dordrecht, Netherlands: Kluwer.

Holling, C. S. 1973. "Resilience and stability of ecological systems." *Annu. Rev. Ecol. Syst.* 4 (1): 1–23.

Holling, C. S., D. W. Schindler, B. W. Walker, and J. Roughgarden. 1995. "Biodiversity in the functioning of ecosystems: An ecological synthesis." In *Biodiversity loss: Economic and ecological issues*, edited by K. G. Mäler, C. Folke, C. S. Holling, and B. O. Jansson, 44–83. Cambridge, UK: Cambridge University Press.

Horne, J. F., and J. E. Orr. 1998. "Assessing behaviors that create resilient organizations." *Employ. Relat. Today* 24 (4): 29–39.

"Industry statement on resilience." Accessed March 2021. https://cdn.ymaws.com/www.nibs.org/resource/resmgr/Docs/Statement_2016-0425.pdf.

Kennett, M., M. Ettouney, P. Schneider, R. F. Walker, and M. Chipley. 2011a. *Integrated rapid visual screening of buildings. Buildings and infrastructure protection series*, edited by M. Kennett. Washington, DC: US Department of Homeland Security.

Kennett, M., M. Ettouney, S. Hughes, R. F. Walker, and E. Letvin. 2011b. *Integrated rapid visual screening of mass transit stations. Buildings and infrastructure protection series*, edited by M. Kennett. Washington, DC: US Department of Homeland Security.

Kennett, M., M. Ettouney, S. Hughes, R. F. Walker, and E. Letvin. 2011c. *Integrated rapid visual screening of tunnels. Buildings and infrastructure protection series*, edited by M. Kennett. Washington, DC: US Department of Homeland Security.

Klein, R. J. T., R. J. Nicholls, and F. Thomalla. 2003. "Resilience to natural hazards: How useful is this concept?" *Environ. Hazards* 5 (1): 35–45.

Love, A. E. H. 1893. *A treatise on the mathematical theory of elasticity.* New York: Dover.

Maguire, B., and P. Hagan. 2007. "Disasters and communities: Understanding social resilience." *Aust. J. Emerg. Manage.* 22 (2): 16–20.

Merriam-Webster Dictionary. 2016. Washington, DC: Library of Congress.

Mileti, D. 1999. *Disasters by design: A reassessment of natural hazards in the United States.* Washington, DC: Joseph Henry Press.

NIAC (National Infrastructure Advisory Council). 2009. *Critical infrastructure resilience final report and recommendations*. Washington, DC: NIAC.

NIST (National Institute of Standards and Technology). 2015. Vol. I of *Community resilience planning guide for buildings and infrastructure systems*. Washington, DC: NIST.

NIST. 2016. Vol. II of *Community resilience planning guide for buildings and infrastructure systems*. Washington, DC: NIST.

NRC (National Research Council). 2010. *Review of the Department of Homeland Security's approach to risk analysis*. Washington, DC: National Academic Press.

NRC. 2012. *Disaster resilience: A national imperative*. Washington, DC: National Academies Press.

NSC (National Security Council). 2011. "Presidential Policy Directive/PPD-8: National preparedness." Accessed May 26, 2018. https://www.dhs.gov/presidential-policy-directive-8-national-preparedness.

NSTC (National Science and Technology Council). 2005. *Grand challenges for disaster reduction—An overview*. Washington, DC: NSTC.

OPS (Office of the Press Secretary). 2013. "Presidential Policy Directive/PPD-21: Critical infrastructure security and resilience." Accessed October 20, 2019. https://www.dhs.gov/sites/default/files/publications/PPD-21-Critical-Infrastructure-and-Resilience-508.pdf.

Paton, D., and D. M. Johnston. 2001. "Disasters and communities: Vulnerability, resilience and preparedness." *Disaster Prev. Manage.* 10 (4): 270–277.

Pimm, S. L. 1984. "The complexity and stability of ecosystems." *Nature* 307 (5949): 321–326.

Rose, A., and S. Y. Liao. 2005. "Modeling regional economic resilience to disasters: A computable general equilibrium analysis of water service disruptions." *J. Reg. Sci.* 45 (1): 75–112.

Russell, S. B. 1898. "Experiments with a new machine for testing materials by impact." *Trans. ASCE* 38: 826.

Timmerman, P. 1981. *Vulnerability, resilience and the collapse of society: A review of models and possible climatic applications*. Toronto: Institute for Environmental Studies, University of Toronto.

UN/ISDR (United Nations/International Strategy for Disaster Reduction). 2002. *Living with risk: A global review of disaster reduction initiatives*. Preliminary version prepared as an interagency effort coordinated by the ISDR Secretariat. Geneva: UN/ISDR.

US Board to Test Iron, Steel and Other Metals. 1879. *Report on a preliminary investigation of the properties of the copper–tin alloys*. Washington, DC: Government Publishing Office.

Walker, B. H., and J. A. Meyers. 2004. "Thresholds in ecological and social–ecological systems: A developing database." *Ecol. Soc.* 9 (2): 3.

Walter, J. 2004. *World disaster report 2004: Focus on community resilience.* Geneva: International Federation of Red Cross and Red Crescent Societies.

Wildavsky, A. 1991. *Searching for safety.* New Brunswick, Canada: Transaction.

Young, T. 1807. *A course of lectures on natural philosophy and the mechanical arts.* London: William Savage.

Zhou, H., J. Wang, J. Wan, and H. Jia. 2010. "Resilience to natural hazards: A geographic perspective." *Nat. Hazards* 53 (1): 21–41.

CHAPTER 2

OBJECTIVE RESILIENCE OF INFRASTRUCTURE SYSTEMS

Amar A. Chaker

2.1 INTRODUCTION

The purpose of this chapter is to review the evolution over the last several decades of the design and performance objectives of civil engineering systems such as structures and transportation or flow networks subjected to various actions (e.g., gravity, wind, snow, ice and earthquake loads, fatigue, corrosion, scour, man-made or accidental loads, and so on) and to provide an overview of recent advances in the objective resilience of infrastructure systems.

This chapter highlights the steady increase in society's expectations in terms of performance objectives, starting from the safety and reliability of a single element exposed to a single or multiple hazards and of the risk associated with a failure, then expanding these requirements in terms of safety, reliability, and risks associated with failures of an entire infrastructure system. These expectations are expanded further to the case of a set of interconnected, interdependent infrastructure systems, and, finally, to the evaluation and management of the resilience of entire infrastructure systems.

2.2 HAZARDS, THREATS, AND DISRUPTIVE OR EXTREME EVENTS

Such civil infrastructure systems are exposed over their lifetime to a range of hazards of different types:

- *Hydrometeorological hazards* such as tropical cyclones called typhoons (Western Pacific Ocean), hurricanes (Eastern Pacific and North Atlantic Oceans) or cyclones (Indian and South Pacific Oceans), extratropical cyclones, tornadoes, flooding, extreme heat, drought, and wildfire.

- *Geotechnical hazards* such as landslide, mudflow, or debris flow.
- *Geophysical hazards* such as earthquake ground shaking often accompanied by secondary effects (landslide, fault rupture, soil liquefaction, rock fall, settlement, seiche or fire following earthquake), tsunami (generated by the displacement of water following submarine earthquakes or in some cases submarine landslides), volcanic eruption, lahar, and geomagnetic storm.
- *Technological hazards* such as accidental explosions, fires, collisions, and release of dangerous materials in the environment (oil spills, radioactive contamination, and so on).
- *Man-made hazards* such as terrorism and cyber-attacks.

These hazards may produce *primary and secondary effects* that can lead to the failure of portions of infrastructure systems. For example, the primary effect of an earthquake is ground shaking, which can cause the failure of bridge structures or the compaction of fill at bridge abutments. Its secondary effects include surface fault rupture, tectonic deformation, ground subsidence, lateral spreading, soil liquefaction, landslides, rock falls, flooding, fire, and even tsunamis. The primary effect of hurricanes is high-velocity winds, which can be accompanied by the hurling of debris and flooding as secondary effects.

These hazards, threats, or disruptive events vary widely by the *length of the warning* before their onset. In some cases, there is no warning (terrorism, explosion) or an extremely short warning (earthquake). In other cases, there is a short warning of a few minutes to a few hours (tsunami). For hazards such as hurricanes, wildland fires, and floods, the warning time may be of a few hours to few days. For failure induced by phenomena such as scour, corrosion, and fatigue, the warning time is usually longer and is typically measured in years. Finally, the effects of sea-level rise and climate change may take years or decades to be noticeable.

The hazards may occur one at a time or simultaneously (*multihazard* situations). For example, hurricane winds are often accompanied by storm surge and heavy precipitations, as in the case of Hurricanes Katrina (August 2005), Harvey (August 2017), Irma (September 2017), and Maria (September 2017). Hurricane (or Superstorm) Sandy added extreme tide to these three threats in October 2012. In the case of Sacramento, California, one of the threat scenarios is the joint occurrence of an earthquake and a Pacific storm that would damage the levees and create the risk of major flooding in the Sacramento-San Joaquin River Delta. Similarly, the Pacific Northwest region of the United States and Canada is exposed to the possibility of strong ground shaking owing to an earthquake on the Cascadia Subduction Zone followed by a tsunami.

Finally, highly improbable events with extreme consequences are "black swans,", such as the 2011 Tohoku Earthquake, a magnitude 9.0 to 9.1 (M_w) undersea megathrust earthquake off the coast of Japan, accompanied by powerful tsunami waves that reached heights of up to 40.5 m and traveled up to 10 km inland and an unexpected combination of events that caused the complete failure of defenses in depth, resulting in the meltdown of three reactors in the Fukushima Daiichi nuclear power complex and the release of radiation over a wide area.

The terminology used when discussing the impact of hazards on infrastructure systems refers to concepts with precise meanings that have now been standardized within the hazards engineering community:

- *Hazard* refers to a threat or to a disruptive or extreme event that may incur damage and losses.
- *Vulnerability* is the likelihood of damage or loss for a given hazard level.
- *Consequences* include injuries, loss of life, damage, economic loss (direct and indirect, including loss of continuity of operations), and people displaced by disasters.
- *Risk* refers to expected losses.
- *Fragility* is the probability of an undesirable outcome as a function of excitation, usually in the context of structural response.

2.3 INFRASTRUCTURE SYSTEMS

Civil infrastructure systems that provide vital services for essential functions of modern societies go by several different names such as lifelines, lifeline networks, and critical infrastructure. They typically are geographically distributed networks or graphs of links and nodes required for delivering energy, communications, water and wastewater, and transportation services and for supporting financial transactions and supply chains. They are physically distributed over a wide area and so is their exposure to hazards. All these systems support the economic well-being, security, and social fabric of the communities they serve.

Failure at one point often means failure of the entire link or even cascading impacts across multiple links. They can also be essential buildings that are a part of the critical infrastructure (government, emergency response, security, telecommunications, power, and so on). The main physical infrastructure systems include the following:

- Transportation (roads, rail, air, and water-borne).
- Telecommunications (wired and wireless) networks.
- Internet.
- Electrical power generation, transport, and distribution.

- Gas and liquid fuels (pipeline networks) storage, transport, and distribution networks.
- Water storage, treatment, transport, and distribution.
- Wastewater collection and treatment systems.
- Stormwater systems; and
- Essential buildings.

Financial transaction systems (credit card, interinstitution payments, stock market, and so on) and supply chains (trucking, warehousing, intermodal transfer of parts and subassemblies) are two other types of infrastructure systems essential to modern societies that are built upon physical infrastructure systems (communication and multimodal transportation networks, respectively), while remaining distinct from them because they provide different functions (Merz et al. 2007).

The infrastructure systems identified here also appear, explicitly or implicitly, in the list of 16 critical infrastructure sectors adopted by the US Dept. of Homeland Security (Cybersecurity and Infrastructure Security Agency 2021). The lists are not identical because the DHS list is organized by functional sectors, whereas the list considered here is organized primarily by physical systems.

2.4 SAFETY AND RELIABILITY OF INFRASTRUCTURE ELEMENTS

Infrastructure systems must maintain their integrity to remain operational and keep providing their intended function. This means that their nodes and links must be reliable and safe and the connectivity within the network such as the ability of structural members to transfer forces or pipe segments to carry fluid must be maintained. The safety and reliability of infrastructure elements and systems are performance objectives that have always been a primary concern of engineers designing and building infrastructure.

Infrastructure elements can, in general, be characterized by a "capacity," such as the height of a protection levee or that of a bridge over a body of water, the load-carrying capacity of a structural member or of a foundation, the traffic capacity of a road segment, the flow capacity of a pipe, and so on. The exact capacity of an infrastructure element is usually an uncertain quantity, but engineers can identify or model the likely range of capacities given multiple factors. For example, the material characteristics of the concrete and of the steel reinforcement in a reinforced concrete beam, the geometry of the cross section, and the position of the steel within the cross section have some uncertainty attached to them, which, in turn, creates uncertainty on the load-carrying capacity of the beam. These infrastructure elements are also subjected to uncertain "demands" such as the maximum height reached by a flood and the load exerted on a structural member by wind, snow, or earthquake ground shaking. Here, the source of uncertainty on the demand is the variability inherent to hydroclimatological or geophysical hazards.

To study the safety of one infrastructure element, engineers address uncertainty by considering the load (or demand) on the element and the resistance (or capacity) of the element as random variables. The most used probabilistic models include models for discrete random trials, models for random occurrences, and models for extreme value distributions (e.g., wind speed, snow loads, or volume of precipitations). A topic of special interest is the distribution of extreme values of random variables such as wind speed, snow loads, volume of precipitations, and so on. The frequency of occurrence of random events such as earthquake ground shaking exceeding a given value is another area of interest.

These probabilistic models can be very useful: Even if the joint probability distribution of demand D and capacity C is not known, Chebyshev's inequality makes it possible to provide a lower bound on the probability of the capacity exceeding the demand, that is, on the reliability of the infrastructure element based only on their second moments or variances σ_C^2 and σ_D^2 (Benjamin and Cornell 1970).

If the margin M is defined as $M = C - D$, the probability of failure is

$$p_f = P(C < D) = P(M < 0) \leq P\big[|M - \mu_M| \geq \mu_M\big] \leq (\sigma_M / \mu_M)^2$$

and the reliability, defined as $1 - p_f$, is at least $1 - (\sigma_M / \mu_M)^2$, or $1 - V_M^2$, where V_M is the coefficient of variation of M.

Early attempts at addressing safety and reliability in engineering design were based on the concept of factor of safety. For example, in allowable stress design (ASD) codes, allowable strength is defined as the nominal strength divided by the safety factor

$$R_{\text{allow}} = R_n / FS$$

and one must check that the required strength is less than or equal to the allowable strength:

$$R_{\text{req}} \leq R_{\text{allow}}$$

Similarly, when examining the stability of a slope, the moment of the forces that tend to move the soil mass is compared with the resisting moment induced by the shear stresses on the assumed slip surface, divided by a factor of safety. Note, however, that the concept of the factor of safety fails to capture the likelihood that the demand on the element might exceed its capacity. Two elements designed with the same factor of safety may have different failure probabilities. For this reason, the focus turns to the probability of failure and the reliability of the infrastructure element:

Probability of failure: $p_f = P(\text{Demand} > \text{Capacity})$ or $P(\text{Load} > \text{Resistance})$

Reliability: $1 - p_f$

For example, if the capacity C and demand D are normal random variables $N(\mu_C, \sigma_C)$ and $N(\mu_D, \sigma_D)$, the margin $M = C - D$ is $N(\mu_M, \sigma_M)$. For statistically independent C and D, $\sigma_M^2 = \sigma_C^2 + \sigma_D^2$, and $M - \mu_M$ is $N(0, 1)$.

The probability of failure is $p_f = P(M \leq 0) = \Phi(-\mu_M/\sigma_M) = 1 - \Phi(\mu_M/\sigma_M)$, where Φ is the cumulative probability distribution function of the normally distributed variable with mean 0 and standard deviation 1. The probability of safety or reliability is $p_s = 1 - p_f = \Phi(\mu_M/\sigma_M)$. If the *safety index* β is defined as

$$\beta = \mu_M/\sigma_M$$

the probability of failure is $p_f = 1 - \Phi(\beta)$ and the reliability is $p_s = \Phi(\beta)$ (Ang and Tang 1975, 1984).

This theory of safety and reliability, expanded to the case of multiple variables, is the rational basis of the load and resistance factor design (LRFD) approach of modern building codes. The LRFD specification accounts separately for the uncertainty of the applied loads using load (i.e., overload) factors applied to the required strength side of the limit state inequalities and for material and construction variabilities through resistance (reduction) factors on the nominal strength side of the limit state inequality. The individual loads are then combined using load combination equations that consider the probability of loads occurring simultaneously.

It is worth noting that the traditional models used for hydrometeorological hazards (extreme wind, floods, precipitations, snow falls, and so on) assume stationarity, that is, the model parameters do not change over time, an assumption that must be revisited when the climate is changing.

2.5 SAFETY AND RELIABILITY OF INFRASTRUCTURE SYSTEMS

Evaluating the safety and reliability of an entire infrastructure system is substantially more involved than analyzing the safety and reliability of an infrastructure element, because there are numerous failure modes for links and nodes of the network that would prevent it from functioning as planned. System analysis and evaluation techniques include event tree diagrams representing a sequence of possible events, each with its probability of occurrence, and leading to a specific consequence.

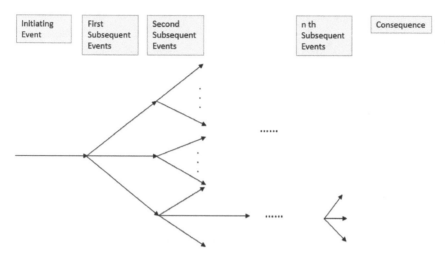

Figure 2-1. Event tree example.

Consequences such as deaths, serious injuries, and economic losses associated with "failure" events that could result from the initiating event and from subsequent events are then evaluated in an event tree. To construct the event tree (Figure 2-1), one must identify different sequences of subsequent events and their consequences. Each path is a unique sequence of events resulting in a specific consequence (Ang and Tang 1975, 1984).

Fault tree analysis can also be used in that context to capture a combination of events that may result in the same outcome.

A quantitative analysis framework to conduct risk analysis can start with a very detailed list of basic events that can occur in the context considered, together with an estimate of their probability of occurrence based on observed data. These events are then combined using AND or OR gates (or operators) to construct a fault tree representing causal relationships, with the main failure event at the top. Each event in the tree may be the outcome of a subtree. The probability of occurrence of an event at a given level is obtained by multiplication (AND gate) or addition (OR gate) of the probabilities of the events at the immediately lower level that may lead to the event considered. Fault tree–based reliability analyses are common for mechanical, industrial, and nuclear engineering systems. Figure 2-2 illustrates the process of constructing the fault tree for the case of a hospital whose electrical power is normally provided by the electrical power system (Lewis 1996). When power is lost, the voltage drop is detected, which triggers the start of an emergency diesel generator. The failure event at the top (power blackout) occurs when off-site power is lost,

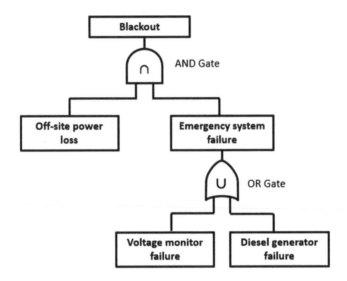

Figure 2-2. Simple fault tree example.

and the emergency power system fails. In turn, the emergency power system fails when the voltage monitor fails, or the diesel generator fails.

By representing the fault tree leading to the failure event horizontally and connecting it to the event tree showing the "failure" branches and their associated consequences, an elegant "bow-tie" representation of system risk analysis can be obtained.

In some elaborate cases, it is useful to develop fragility curves for elements, components, and subsystems that describe the probability of an undesirable outcome (such as amount of loss or reaching a specific performance level in performance-based earthquake engineering) as a function of a demand that is a random variable. In extremely complex cases, systems engineering techniques, graph theory, and discrete event simulation are used.

The typical strategies for enhancing the reliability of such systems include the following:

- Using fusible or sacrificial elements and easily replaceable links, for example, disposable knee in steel-braced frames.
- Incorporating redundancy, for example, combination of moment-resisting frames, braced frames, core, and walls. Transportation systems usually have redundancy so that, most often, a variety of transportation options and alternate routes are available when components of the network are damaged.
- Providing multiple lines of defense, for example, Hurricane Sandy in 2012 inundated all three airports that serve New York City, crippling travel for days. St. Paul Downtown Airport in Minnesota, which has

been frequently flooded by the Mississippi, now has a portable flood wall that can be erected if the river starts to overflow. With the help of a $28 million federal grant, La Guardia Airport in New York is adding a flood wall, rainwater pumps, and a new drainage system for the airfield, as well as upgrading its emergency electrical substations and generators.

- Providing standby components that can be deployed to replace failing ones, for example, foldable bridge; standby generators.
- Providing an alternate or substitute solution, for example, in some cases, it may be possible to rely on the substitution of one mode of transportation for another. Following the span collapse on the San Francisco–Oakland Bay Bridge during the 1989 Loma Prieta earthquake, many motorists used the Bay Area Rapid Transit and the ferry.
- Subdividing the entire infrastructure system into several autonomous systems, so that failure affects only a portion of the system that will operate in degraded mode, for example, microgrids for electrical power generation, transport, and distribution networks.

An example of a complex risk assessment study for the hurricane protection system of New Orleans conducted by the Interagency Performance Evaluation Task Force for the US Army Corps of Engineers is shown schematically in Figure 2-3 (USACE 2009).

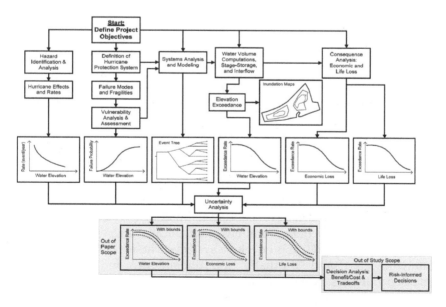

Figure 2-3. Schematic diagram of the hurricane protection system of New Orleans.

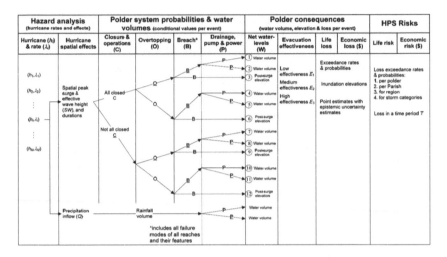

Figure 2-4. Event tree model for the New Orleans hurricane protection system risk analysis.

The event tree model developed for the New Orleans hurricane protection system risk analysis is shown in Figure 2-4 (USACE 2009).

2.6 SAFETY AND RELIABILITY OF A SET OF INTERCONNECTED AND INTERDEPENDENT INFRASTRUCTURE SYSTEMS

The safety and reliability of infrastructure systems can be studied for an individual, isolated infrastructure system as presented in the previous section. However, such systems are usually connected to, and mutually dependent on, other infrastructure systems: For example, the water, wastewater, and highway transportation and telecommunication systems of a city can be severely affected when electrical power is not available to operate pumps, traffic lights, telephone switches, and so on. Such interconnected, interdependent systems are usually present at the scale of a city or region, as schematically represented in Figure 2-5 (Pederson et al. 2006) and Figure 2-6 (Mieler and Mitrani-Reiser 2018).

The multiple interconnections are depicted in a detailed manner in Figure 2-7 (Peerenboom 2001).

Similar situations can also be found in buildings, which usually have structural, electrical, communications, mechanical, water, wastewater, gas, fire protection, HVAC, and envelope systems. These systems are highly interconnected and mutually interdependent, and the mutual interdependencies create vulnerabilities. A disruption in one infrastructure system can have ripple effects onto other infrastructure systems (failure propagation and performance degradation). A frequent example is when

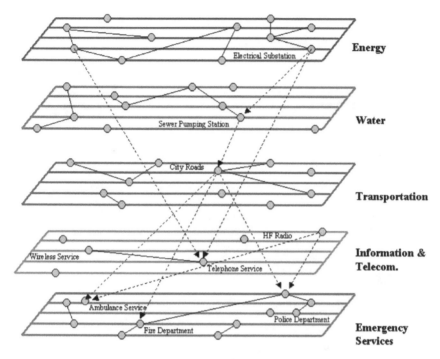

Figure 2-5. Interconnected interdependent infrastructure systems.

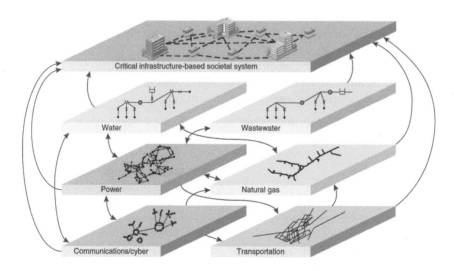

Figure 2-6. Interconnected and mutually interdependent community-level systems.

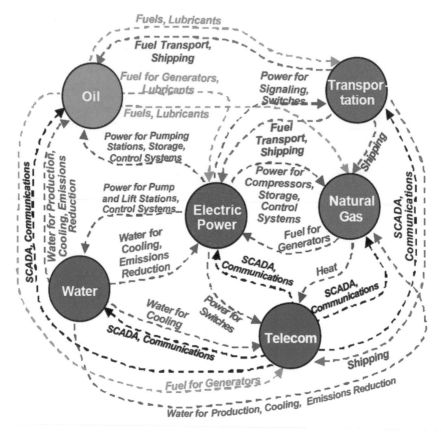

Figure 2-7. Multiply connected interdependent systems.

loss of electrical power can prevent the water distribution, HVAC, and
communication systems of a building from operating.

The interdependencies can be of different types:

- *Physical* if the output of one system is used by another (e.g., the
  system requires electricity to operate).
- *Cyber* (electronic information and control systems).
- *Geographic* (e.g., utilities collocated within the right-of-way of a road
  or carried by a bridge). Utilities, including pipelines, often follow
  transportation infrastructure rights-of-way and cross valleys using
  bridges of the road system. Because of the physical proximity of these
  lifeline systems and of their functional interdependency, interaction
  among them can be significant.
- *Logical* (interdependent via a mechanism that is not a physical,
  cyber, or geographic connection, such as linkage through financial
  systems).

Failures can be of different categories:

Cascading: A disruption in which one infrastructure causes a disruption in a second. Two often overlooked types of cascading impact are the disruption of the supply chain and the displacement and relocation of individuals (which occurred following Hurricane Katrina). Striking examples of cascading impacts include the following:

- The August 14, 2003, North America power blackout shut down water treatment plants and pumping stations.
- Multiple instances of cascading failures occurred during the 2012 Hurricane (or Superstorm) Sandy and the 1994 Northridge earthquake.
- Because of the September 11, 2001, World Trade Center attack, water from ruptured underground pipelines flooded the tunnels beneath the Hudson River and the cable vault of the Verizon building.

Escalating: A disruption in one infrastructure that exacerbates an independent disruption of a second infrastructure. The time for restoration of a failure of a water pipeline can increase because the transportation infrastructure has a failure that prevents parts or repair workers from reaching the failed pipeline.

Common cause: A disruption in two or more infrastructure systems is the result of a common cause. Hurricane Katrina caused the loss of electrical power, natural gas supply, water supply, and telecommunications.

An example of downstream infrastructure dependencies is provided by the 2004 Atlantic hurricane season in Florida (Bigger et al. 2009). Trees that fall on power lines and bring them to the ground, as well as flooding, can cause an interruption of electrical power supply, which itself can affect the fuel delivery system, the communications infrastructure (cell phone service and telephone switchboards), the water/wastewater infrastructure (inoperable pumping systems), and the transportation system (loss of traffic signals). These are examples of first-level dependencies.

The Children's Hospital in New Orleans is located next to the Mississippi River in Orleans Parish in an area that suffered significant long-term flooding because of the breached levee system. However, because it sits on high ground right next to the river, the hospital did not flood after the August 2005 Hurricane Katrina. Nevertheless, the hospital had to shut down and evacuate the patients because of a lack of municipal water for the HVAC/chillers that provide air conditioning for the operating theater, an unexpected second-order downstream dependency (Hapij 2011).

The Tohoku earthquake and tsunami offers multiple examples of higher-order downstream dependencies (Chock et al. 2013, Ewing et al.

2013, Tang 2017). On March 11, 2011, Japan experienced the most powerful earthquake in its history (moment magnitude M_w 9.0), which triggered just over an hour later a 30 ft (10 m) tsunami that washed through the Fukushima Daiichi nuclear power station. The earthquake ground shaking cut the plant off from outside sources of electricity. The tsunami easily topped the plant's sea walls, flooding the underground bunkers containing its emergency generators, cutting off the power supply to the cooling system of the reactors, and causing a meltdown of the core of three of the six reactors and spreading radiation across the region, contaminating large volumes of soil and water, rendering a large zone uninhabitable, and forcing the evacuation of the population. Storing and disposing of the contaminated soil and the contaminated water that has seeped into the plant or has been used to cool spent fuel is a major issue.

The arrival of Superstorm/Hurricane Sandy on October 29, 2012, occurred almost exactly with high tide on the Atlantic Ocean, at the peak of its monthly cycle. The surge, the flooding, and the waves that came with it caused major disruptions in and around New York City (The City of New York 2013). Disruptions to the city's electric utilities, including the inundation of several substations in Southern Manhattan, caused nearly 2 million people to lose power during the storm. In many cases, floodwaters filled basements of buildings and made electrical and mechanical building equipment inoperable. The storm inundated four of the Manhattan steam system six plants. Many highways, roads, railroads, and airports flooded, as did all the six East River subway tunnels. The loss of electrical power made it difficult to pump out tunnels and exacerbated the flooding. The storm also shut down refineries for several weeks, stopped marine and pipeline fuel deliveries for several days, and damaged storage terminals, thus affecting fuel supply for nearly a month. Here again, downstream dependencies and higher-order dependencies can be observed.

Analyzing the safety and reliability of such groups of interconnected and mutually interdependent infrastructure systems requires careful identification and documentation of the interdependencies, the likelihood of failures of infrastructure elements and systems under a range of hazards, and their impact on the functionality of interdependent infrastructure systems. In theory, event and fault tree analyses could still be considered, but because of the huge increase in the number of combinations of fault events to examine across multiple infrastructure systems, such approaches may not be practical. Discrete event simulation allows a more direct approach at the level of the infrastructure networks. It can accurately represent the complexity of the systems and their interdependencies, is now widely available and affordable, and may be the only practical avenue for quantitative analysis.

The approaches discussed here for uncertainty quantification and risk analysis serve as the rational basis for making informed decisions regarding engineering design, retrofit, and investments and risk transfer (through insurance/reinsurance).

2.7 CONCEPT OF RESILIENCE

For the last four decades, many organizations such as government agencies, development banks, universities, and professional organizations have been addressing issues such as disaster preparedness, disaster prevention, vulnerability reduction, continuity of operations, disaster mitigation, disaster risk reduction, and disaster risk management. The case for such efforts is presented very clearly and convincingly in the publication of The International Bank for Reconstruction and Development and The World Bank, *Natural Hazards, Unnatural Disasters* (World Bank 2010). Resilience has become a very common term that is often inserted into proposals to help attract funding. However, the term is also used as part of a conscious effort to avoid the negative connotation of vulnerability and highlight the more positive idea of resilience. The effort to provide a more constructive presentation of disaster risk reduction also increased the use of the word resilience.

In the context of complex interconnected and interdependent infrastructure systems, the *definition of resilience* refers to a new performance objective that implies an expectation of minimal disruption and rapid recovery and restoration of function (Bruneau et al. 2003).

Because of the interconnectedness and interdependency, resilience calls for systems thinking. Resilience goes beyond the evaluation of safety and reliability of such systems and the estimation of direct losses resulting from failures (risk assessment). Resilience values the importance of maintaining the continuity of operations and the indirect cost of the loss of function and highlights the need to promptly restore functionality (bouncing back). This, in turn, implies robustness, by calling for the smallest possible reduction of the level of functionality owing to the disruption, and rapidity by calling for a recovery that is as prompt as possible, which may become possible thanks to redundancy and resourcefulness. Resilience is often summarized and captured using the "4Rs" shortcut as shown in Table 2-1 (Bruneau and Reinhorn 2006).

Moreover, resilience frequently considers the possibility of building back better ("bounce forward" instead of "bounce back"), as illustrated in Figure 2-8, which represents functionality versus time and shows the loss of functionality when a disaster occurs (limited, thanks to the *robustness*), as well as the recovery time. Incorporating *redundancy* and calling upon *resourcefulness* contribute to the *rapidity* of the recovery.

More recent efforts to address the issue in a scientific, rigorous approach using quantitative models and rigorous approaches to evaluate, quantify,

Table 2-1. Four Rs of Resilience.

Attribute	Description
Robustness	Ability to withstand a shock
Redundancy	Functional diversity
Resourcefulness	Ability to mobilize when threatened
Rapidity	Ability to contain losses and recover in a timely manner

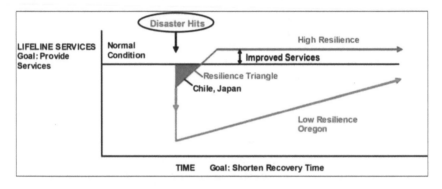

Figure 2-8. Diagram of functionality versus time.
Source: Oregon Seismic Safety Policy Advisory Commission, The Oregon Resilience Plan, February 2013.

and manage resilience use the term "Objective Resilience." These efforts also recognize that climate is changing but cannot be projected with a high degree of certainty, and that climate change introduces nonstationary.

2.8 EVALUATING AND MEASURING RESILIENCE

2.8.1 Screening and Prioritization

Some of the earliest efforts to evaluate and measure resilience of critical infrastructure facilities include the Integrated Rapid Visual Screening Series (IRVS) for buildings, mass transit stations, and tunnels developed by the Department of Homeland Security Science and Technology Directorate in its Buildings and Infrastructure Protection Series (BIPS).

BIPS 04 presents a methodology based on the FEMA 452 publication (FEMA 2005) for the tiered quantifiable assessment of the risk and resilience of all conventional building types with respect to terrorist attacks and selected natural hazards (DHS 2011c). The assessment considers possible interactions among multiple hazards. The IRVS results enable the formulation of action plans and programs to enhance resilience, reduce vulnerability, deter threats, and mitigate potential consequences.

BIPS 02 presents a methodology for quantifiably assessing the risk and resilience of mass transit stations (heavy rail, light rail, commuter rail, trolley, and bus) to terrorist attacks and selected natural hazards (DHS 2011a). Mass transit stations are hubs that allow passengers to board and disembark from mass transit vehicles and to transfer between modes of transport. The primary purpose of the IRVS of mass transit stations is to rank the risk in a group of mass transit stations in a community.

BIPS 03 focuses on a methodology for quantifiably assessing the risk and resilience of tunnels to terrorist attacks and selected natural hazards (DHS 2011b). Many tunnels have limited alternate routes because of geographic constraints or because they are underwater. They can receive collateral damage from attacks on other targets because of their underground or underwater location. The primary purpose of the IRVS of tunnels is to rank the risk in a group of tunnels in a transportation system or region.

2.8.2 Scenarios—Case Studies

Other researchers have used scenarios to assess the impact of disruptive events on the operability of the infrastructure network or system. For example, the seismic vulnerability of the Memphis, TN highway system following a scenario earthquake on the nearby New Madrid fault zone was studied by Werner et al. (2000a, b, 2004) by considering the likely ground shaking expected at the location of all the bridges in the area, the local soil conditions at the bridge sites, the seismic fragility of these bridges, and determining whether each bridge was expected to remain useable, and finally evaluating the likely impact on the connectivity of the highway network.

Similarly, a ShakeOut scenario analysis of the seismic risks to the Southern California highway system was conducted as part of a larger "Golden Guardian" exercise assessing the impact of a magnitude 7.8 earthquake (Werner et al. 2006, 2008). The analysis estimated post-earthquake traffic flows, travel times, and travel demand, as well as the cost of disaster-induced traffic, travel disruption, and repair costs. The results included the effects of ground-shaking hazards and permanent ground displacements, damage repair and traffic disruption estimates, and economic losses.

Wang and Chaker (2004, 2005) conducted a multihazard study of the Columbia River transportation corridor, a significant east–west transportation artery for the Pacific Northwest that includes US Interstate Highway 84, two transcontinental rail lines, inland water navigation on the Columbia River, a navigational lock for river commerce, major electric power and gas lines, lines of communication, and three large hydroelectric facilities. This transportation corridor is exposed to several geologic hazards: landslides along the steep Columbia River Gorge; shallow, crustal earthquakes and Cascadia Subduction Zone earthquakes, accompanied by seiches, landslides, lateral spreading, liquefaction, and fault rupture; and volcanic eruptions and lahars in the Cascade Range. Two low-probability, worst-case scenarios were examined.

A mega landslide on the Cascade Landslide Complex at the toe of which the Bonneville Project infrastructure is located could result in a complete disruption of transportation through the Columbia River Gorge, and heavy damage to, and possibly a complete destruction of, major population areas and facilities downstream and upstream, including low-lying parts of Portland, small cities, dams, and commercial and industrial sites. The second scenario is a dam failure and water release from the John Day Lock and Dam or other major water storage dams upstream in the Columbia basin, owing to a strong earthquake on a nearby fault. Significant damage owing to overtopping of dikes and levees would probably cause disruption to cities, including significant portions of downtown Portland, The Dalles, other smaller cities, power generation facilities, and the transportation infrastructure, such as the Port of Portland, the Portland International Airport, and the Union Pacific terminals.

The severe, long-lasting impact of these geologic hazards on the economy of Oregon affects productive capacity and slows the pace of economic growth and development, leading the authors to call for additional studies and mitigation to better evaluate the risks and lower the vulnerability of this major transportation corridor.

Such models allow performing "what-if" analyses and can be very helpful for studying functionality restoration scenarios and evaluating alternate mitigation options.

2.9 RESILIENCE MANAGEMENT

Several cities, metro areas, regions, and states that are exposed to major hazards have developed resilience plans, the first step in managing resilience. In fact, they have attempted to chart a path to increase resilience over a planning horizon.

The Oregon Resilience Plan (Oregon Seismic Safety Policy Advisory Commission 2013) aims at protecting citizens from life-threatening physical harm by recommending risk-reduction measures and predisaster planning that will allow communities to recover more quickly and with less continuing vulnerability following a Cascadia subduction zone earthquake and tsunami.

The plan's recommendations highlight ways to close the gap that separates expected and desired performance. The plan examines the likely impacts of a magnitude 9.0 Cascadia earthquake and tsunami and estimates the time required to restore functions in each sector if the earthquake were to strike under present conditions. It reviews business and workforce continuity, coastal communities, critical and essential buildings, transportation, energy, information and communications, and water and wastewater systems. The plan defines acceptable time frames to restore functions after a future Cascadia earthquake to fulfill expected resilient performance and recommends changes in practice and policies that will allow Oregon to reach the desired resilience targets.

The Resilient Washington State Subcommittee of the Washington State Emergency Management Council's Seismic Safety Committee prepared the November 2012 report titled *A Framework for Minimizing Loss and Improving Statewide Recovery after an Earthquake* which provides the following:

- General assessment of the current recovery capacity of the state's major systems and infrastructure, including estimates of the time it is likely to take for each component to recover following a serious earthquake.
- Target time frame for each component—that is, the time frame within which a given component *ought* to recover to ensure that the state is resilient.
- Top 10 recommendations for improving statewide resilience.
- Summary of interdependencies.

Similarly, the Association of Bay Area Governments Resilience Program (California) has produced the report *Regional Resilience Initiative—Policy Agenda for Recovery* (ABAG 2013) and numerous publications and guides (ABAG 2021a), and the report *Final Plan Bay Area 2050* (ABAG 2021b). The report spells out goals in the areas of housing, the economy, transportation, and the environment across the Bay Area's nine counties. The San Francisco Planning and Urban Research Association (SPUR) prepared numerous documents for the Resilient City project in San Francisco (San Francisco Planning and Urban Research Association 2021).

Following the devastating effects of Hurricane Sandy in the fall of 2012, New York City launched the "New York City Special Initiative for Rebuilding and Resiliency" to produce a plan to provide additional protection for New York's infrastructure, buildings, and communities from the impacts of climate change. That initiative generated the report titled *A Stronger, More Resilient New York* (New York City Special Initiative for Rebuilding and Resiliency 2013.), which is a road map for producing a truly resilient city. The comprehensive report addressed the key components of infrastructure (coastal protection, buildings, utilities, liquid fuels, telecommunications, transportation, water and wastewater, environmental protection, and remediation, health care, and parks), as well as community and economic issues (preparedness and response, economic recovery, and insurance).

A small number of recent efforts have been attempted to adopt a more quantitative approach to resilience management (Nasrazadani and Mahsuli 2017). For example, Shahandashti and Pudasaini (2019) have developed a network-level seismic vulnerability assessment model to identify critical pipes for the proactive seismic rehabilitation of the water pipe network when, as is always the case, budgets are limited.

2.10 STRATEGIES FOR PROVIDING AND ENHANCING RESILIENCE

Strategies for resilience include those that *enhance reliability*, discussed in the section on Safety and Reliability of Infrastructure Systems, and those that *accelerate recovery*.

Thinking outside of the box can lead to innovative ideas. For example flood protection can be provided by structural strategies such as movable storm gates (Rotterdam, Hamburg, London, Venice) or nonstructural strategies, such as building with nature and domesticating natural forces like wind and water and using natural material such as sand and vegetation, to hold back the sea (The World Bank 2017, 2019, Bridges et al. 2018). Green infrastructure can be an alternate mitigation option.

Financial tools for risk transfer such as insurance, reinsurance, and cat bonds can play a role in blunting the setback and enabling a faster recovery. However, they do not eliminate the losses and the risk.

2.11 CONCLUSION

Over the past several decades, society has had increasing expectations, starting from the safety and reliability of a single element, the capacity of which is a random variable, through the reliability of a complex infrastructure network, and then the reliability of a complex system of interconnected and interdependent infrastructure networks, and finally, the resilience of a complex system of interconnected and interdependent infrastructure networks in a multiple hazards context.

The body of knowledge to understand objective infrastructure resilience has kept pace with the increase in societal expectations, as shown by the efforts to advance knowledge undertaken by ASCE and other organizations. They include the large body of literature published by ASCE in journals (*Natural Hazards Review*, *ASCE-ASME J. of Risk and Uncertainty*

Analysis—Part A Civil Engineering, J. of Infrastructure Systems, J. of Structural Engineering, J. of Geotechnical and Geo-environmental Engineering, J. of Engineering Mechanics, J. of Hydraulic Engineering, J. of Hydrologic Engineering), conference proceedings, standards, manuals of practice, committee reports, and books. Other signs of the interest in resilience are the activities of numerous ASCE entities: Infrastructure Resilience Division, other resilience-related committees within the nine Institutes of ASCE, including the EMI Objective Resilience Committee, postdisaster investigations teams, and so on.

2.12 RECOMMENDATIONS

- In addition to resilient physical infrastructure, people are also needed for infrastructure to function. For example, after Hurricane Sandy, many people could not get to work, which prevented numerous infrastructure systems from being operational. Therefore, resilience analyses must include people.
- Community resilience requires resilience of the physical infrastructure systems, people, and the organizations that manage the infrastructure. It is addressed in other chapters of this Manual of Practice.
- An emerging concern is that of cyber security, as the systems that control access to the command and operations that are ubiquitous in modern civil infrastructure systems have created a vulnerability that must be addressed.
- Knowledge gaps remain and point to possible areas of research that require urgent attention. They include the following:
  - Early detection of impending structural failure (e.g., detection of the incipient collapse of a building).
  - Design of building structures against tornados.
  - Tsunamis' forces on structures.
  - Bioinspired self-healing materials.
  - Damage-tolerant architected materials and more generally, resilience-enhancing materials.
  - Application of systems engineering techniques to complex systems of interconnected and interdependent infrastructure networks in a multiple hazards context.
  - "Digital twins" of smart cities/communities using models, sensors, communications, and Internet to understand how they would respond and evolve over time to various disruptive scenarios or causes and make them more resilient.
- Challenges posed by climate change to infrastructure resilience.

REFERENCES

ABAG (Association of Bay Area Governments). 2013. "Regional resilience initiative—Policy agenda for recovery." Accessed October 3, 2021. https://abag.ca.gov/sites/default/files/regional_resilience_initiative_policy_plan_march_2013.pdf.

ABAG. 2021a. "Tools and resources." Accessed October 3, 2021. https://abag.ca.gov/tools-resources.

ABAG. 2021b. "Plan Bay Area 2050." Accessed November 7, 2021. https://www.planbayarea.org/sites/default/files/documents/Plan_Bay_Area_2050_October_2021.pdf

Ang, A. H.-S., and W. H. Tang. 1975. *Probability concepts in engineering planning and design, Vol. I—Basic principles*. Hoboken, NJ: Wiley.

Ang, A. H.-S., and W. H. Tang. 1984. *Concepts in engineering planning and design, Vol. II—Decision, risk, and reliability*. Hoboken, NJ: Wiley.

Benjamin, J. R., and C. A. Cornell. 1970. *Probability, statistics and decision for civil engineers*. New York: McGraw Hill.

Bigger, J. E., M. G. Willingham, F. Krimgold, and L. Mili. 2009. "Consequences of critical infrastructure interdependencies: Lessons from the 2004 hurricane season in Florida." *Int. J. Crit. Infrastruct.* 5 (3): 199–219.

Bridges, T. S., E. M. Bourne, J. K. King, H. K. Kuzmitski, E. B. Moynihan, and B. C. Suedel. 2018. *Engineering with nature: An Atlas*. ERDC/EL SR-18-8. Vicksburg, MS: US Army Engineer Research and Development Center.

Bruneau, M., S. Chang, R. Eguchi, G. Lee, T. O'Rourke, A. Reinhorn, et al. 2003. "A framework to quantitatively assess and enhance the seismic resilience of communities." *Earthquake Spectra* 19 (4): 733–752.

Bruneau, M., and A. Reinhorn. 2006. "Overview of the resilience concept." In *Proc., 8th U.S. National Conf. on Earthquake Engineering*, 3168–3176. Oakland, CA: Earthquake Engineering Research Institute.

Chock, G., I. Robertson, D. Kriebel, M. Francis, and I. Nistor. 2013. *Tohoku, Japan, earthquake and tsunami of 2011: Performance of structures under tsunami loads*. New York: ASCE.

Cybersecurity and Infrastructure Security Agency. 2021. "Critical infrastructure sectors." Accessed October 3, 2021. https://www.cisa.gov/critical-infrastructure-sectors

DHS (Department of Homeland Security). 2011a. *Integrated rapid visual screening of mass transit stations*. Buildings and Infrastructure Protection Series, BIPS 02. Washington, DC: DHS.

DHS. 2011b. *Integrated rapid visual screening of tunnels*. Buildings and Infrastructure Protection Series, BIPS 03. Washington, DC: DHS.

DHS. 2011c. *Integrated rapid visual screening of buildings*. Buildings and Infrastructure Protection Series, BIPS 04. Washington, DC: DHS.

Ewing, L., S. Takahashi, and C. M. Petroff et al. 2013. *Tohoku, Japan, earthquake and tsunami of 2011—Survey of coastal structures.* New York: ASCE.

FEMA. 2005. *Assessment: A how-to guide to mitigate potential terrorist attack against buildings.* FEMA 452. Washington, DC: FEMA.

Hapij, A., ed. 2011. *Multidisciplinary assessment of critical facility response to natural disasters—The case of Hurricane Katrina.* New York: ASCE.

Lewis, E. E. 1996. *Introduction to reliability engineering.* 2nd ed. Hoboken, NJ: Wiley.

Merz, M., M. Hiete, V. Bertsch, and O. Rentz. 2007. "Decision support for managing interruptions in industrial supply chains." In *Forum DKKV/ CEDIM: Disaster reduction in climate change.* Karlsruhe: Center for Disaster Management and Risk Reduction Technology, Karlsruhe Institute of Technology.

Mieler, M. W., and J. Mitrani-Reiser. 2018. "Review of the state of the art in assessing earthquake-induced loss of functionality in buildings." *J. Struct. Eng.* 144 (3): 04017218.

Nasrazadani, H., and M. Mahsuli. 2017. "Probabilistic quantification of community resilience using discrete event simulation, COMPDYN 2017." In *Proc., 6th ECCOMAS Thematic Conf. on Computational Methods in Structural Dynamics and Earthquake Engineering*, edited by M. Papadrakakis and M. Fragiadakis, 1994–2004. https://2017.compdyn. org/

New York City Special Initiative for Rebuilding and Resiliency. 2013. "A stronger, more resilient New York." Accessed October 3, 2021. https://www1.nyc.gov/site/sirr/report/report.page.

Oregon Seismic Safety Policy Advisory Commission. 2013. *The Oregon resilience plan.* Salem, OR: Oregon Seismic Safety Policy Advisory Commission.

Pederson, P., D. Dudenhoeffer, S. Hartley, and M. Permann. 2006. *Critical infrastructure interdependency modeling: A survey of U.S. and international research.* Idaho Falls, ID: Idaho National Laboratory.

Peerenboom, J. 2001. *Infrastructure interdependencies: Overview of concepts and terminology.* Argonne, IL: Infrastructure Assurance Center, Argonne National Laboratory.

Resilient Washington State Subcommittee of the Washington State Emergency Management Council's Seismic Safety Committee. 2012. *A framework for minimizing loss and improving statewide recovery after an earthquake.*

San Francisco Planning and Urban Research Association. 2021. "The resilient city." Accessed October 3, 2021. https://www.spur.org/featured-project/ resilient-city?utm_medium=redirect&utm_source=resilient_city.

Shahandashti, S. M., and B. Pudasaini. 2019. "Proactive seismic rehabilitation decision-making for water pipe networks using simulated annealing." *Nat. Hazard. Rev.* 20 (2): 04019003.

Tang, A., ed. 2017. *Tohoku, Japan, earthquake and tsunami of 2011—Lifeline performance.* New York: ASCE.

USACE (US Army Corps of Engineers). 2009. *Performance evaluation of the New Orleans and Southeast Louisiana hurricane protection system.* Final Report of the Interagency Performance Evaluation Task Force, Volume VIII—Engineering and Operational Risk and Reliability Analysis. Washington, DC: USACE.

Wang, Y., and A. Chaker. 2004. *Geologic hazard study for the Columbia River Transportation Corridor.* Open File Rep. OFR O-04-08. Portland, OR: State of Oregon Dept. of Geology and Mineral Industries.

Wang, Y., and A. Chaker. 2005. "A Preliminary Study of Geologic Hazards for the Columbia River Transportation Corridor." In Vol. 2 in *Acceptable risk processes,* edited by C. Taylor and E. VanMarcke, 194–223. New York: ASCE.

Werner, S. D., S. Cho, and R. T. Eguchi. 2008. *The ShakeOut scenario supplemental study : Analysis of risks to Southern California Highway system.* Washington, DC: USGS.

Werner, S. D., S. Cho, C. E. Taylor, J. P. Lavoie, C. K. Huyck, and R. T. Eguchi. 2006. *REDARS2 demonstration project for seismic risk analysis of highway systems.* Sacramento, CA: California Dept. of Transportation.

Werner, S. D., C. E. Taylor, S. Cho, J. P. Lavoie, C. K. Huyck, C. Eitzel et al. 2004. "New developments in seismic risk analysis of highway systems." In *Proc., 13th World Conf. on Earthquake Engineering,* Paper No. 2189.

Werner, S. D., C. E. Taylor, and J. E. I. I. Moore. 2000a. "New developments in seismic risk analysis of highway systems." In *Proc., 12th World Conf. on Earthquake Engineering.*

Werner, S. D., C. E. Taylor, J. E. I. I. Moore, J. S. Walton, and S. Cho. 2000b. *A Risk-based methodology for assessing the seismic performance of highway systems.* Tech. Rep. MCEER-00-0014. Buffalo, New York: Multidisciplinary Center for Earthquake Engineering Research. http://www.buffalo.du/mceer/catalog.host.html/content/shared/www/mceer/publications/MCEER-00-0014.detail.html.

World Bank. 2010. "Natural hazards, unnatural disasters: The economics of effective prevention." Accessed October 5, 2021. https://openknowledge.worldbank.org/handle/10986/2512.

World Bank. 2017. *Implementing nature-based flood protection—Principles and implementation guidance.* Washington, DC: World Bank Group.

World Bank. 2019. "Nature-based solutions for disaster risk management, Vol. 2: Fact sheet." Accessed September 24, 2021. http://documents.worldbank.org/curated/en/908411551126569861/Fact-Sheet.

CHAPTER 3

ACHIEVING OPERATIONAL RESILIENCE THROUGH CODES, STANDARDS, METRICS, AND BENCHMARKS

Ryan M. Colker

3.1 INTRODUCTION

In 2005, the National Institute of Building Sciences (NIBS), at the direction of the US Congress, undertook an examination of Federal Emergency Management Agency (FEMA) grants to determine the benefits derived from investments in mitigation against natural hazards. The study produced the oft-cited statistic that $1 spent on hazard mitigation derived $4 in benefit (NIBS 2005). In 2018, NIBS released the first interim report in a series of updates examining mitigation strategies beyond FEMA grant programs. In its 2019 Interim Report, NIBS found that (among other things) the adoption of up-to-date (2018) building codes compared with those present around 1990 provides an $11 benefit for every $1 invested (NIBS 2018) (see Table 3-1).

In addition to the NIBS study, building codes have been recognized as a valuable component of hazard mitigation. Building codes serve as the foundation for community resilience through their role in disaster risk reduction (ICC and ANCR 2018). A FEMA analysis estimated approximately $500 million in annualized losses avoided in eight southeastern states because of the adoption of modern building codes (FEMA 2014). Effective and well-enforced building codes in Missouri have reduced hail damage to homes by 10% to 20% on average (Czajkowski and Simmons 2014). And, in the ten years following Florida's adoption of a statewide building code, the code's adoption and application reduced windstorm actual losses by as much as 72%, producing $6 in reduced loss to $1 of added cost (Simmons 2018).

The US Congress through passage of the Disaster Recovery Reform Act of 2018 (DRRA) authorized FEMA to assist states in the adoption and

49

Table 3-1. Benefit–Cost Ratios for Various Mitigation Investments.

	ADOPT CODE	ABOVE CODE	BUILDING RETROFIT	LIFELINE RETROFIT	FEDERAL GRANTS
National Institute of BUILDING SCIENCES — Overall Benefit-Cost Ratio	11:1	4:1	4:1	4:1	6:1
Cost ($ billion)	$1/year	$4/year	$520	$0.6	$27
Benefit ($ billion)	$13/year	$16/year	$2200	$2.5	$160
Riverine Flood	6:1	5:1	6:1	8:1	7:1
Hurricane Surge	not applicable	7:1	not applicable	not applicable	not applicable
Wind	10:1	5:1	6:1	7:1	5:1
Earthquake	12:1	4:1	13:1	3:1	3:1
Wildland-Urban Interface Fire	not applicable	4:1	2:1	not applicable	3:1

Copyright © 2019 The National Institute of Building Sciences

enforcement of the latest codes. Federal agencies have also recognized the impacts. FEMA funding for restoration of public facilities, known as Public Assistance funding, has historically honored funding requirements for satisfying the locally adopted codes. FEMA's strategic plan for 2018 to 2022 includes a goal to build a culture of preparedness and includes building codes as a key component in meeting this objective (FEMA 2018). The Mitigation Framework Leadership Group (MitFLG), an effort of 14 federal government agencies and representatives of state, local, and tribal governments, released a National Mitigation Investment Strategy (NMIS) to facilitate coordination in mitigation investments. The NMIS identifies building codes as an essential strategy in achieving investment goals. Recommendation 3.1 encourages communities to adopt and enforce up-to-date building codes and recommends that up-to-date codes be required for all federal and state grants and programs (MitFLG 2019).

Despite these significant benefits, the code requirements at the state and local levels vary significantly. Many jurisdictions across the country have either not adopted building codes or not kept up with adopting newer editions. Many states set code baselines statewide, whereas in 21 states, local governments determine what, if any, building codes apply in their jurisdictions. Currently, five states representing 12% of the population have state building codes that are nine or more years old and, where local governments determine code adoption, upward of 25% and 10% of residents in some Midwest and Gulf Coast states, respectively, use codes that are just as dated.

Regardless of whether a community has a code in place or is on an older edition of a code, designers and owners of individual projects can still capture many of the benefits contained in recent editions of the model codes. Building codes provide objective criteria for the design and construction of buildings that can be applied both communitywide and at the individual project level. Codes and standards are one of the tools engineers and designers use to fulfill their obligations to protect health, safety, and welfare. Compliance with the content of codes and standards provides an objective basis to support resilience but as discussed in greater detail in the following, community-level resilience requires building on top of the strong base that codes provide. Building codes are necessary but not sufficient to the achievement of a resilient community. In addition, the lack of codes or overly old codes within a jurisdiction does not release a designer's obligation to deliver a safe, sustainable, and resilient project—the standard of care may extend beyond minimum standards, opening up designers to liability claims (CLF 2018).

While this chapter focuses primarily on codes and standards that support resilient buildings, similar sets of criteria exist in other infrastructure systems—many of them with ASCE involvement. The general principles discussed apply across all these systems—particularly when using a whole community perspective.

3.1.1 Definitions

Numerous scholarly papers have parsed the myriad existing definitions for resilience (Meerow et al. 2016, Haimes 2009, Rose 2004). Such reviews often point back to the initial work of Holling in the early 1970s in the field of ecology (Holling 1973). In addition to ecology, the concept of resilience is present in the fields of psychology, business, material science, and others. Fortunately, no matter the discipline, the general concept of resilience remains the same, as discussed in Chapter 1: reducing impacts in the face of adversity and recovering function within a reasonable time. This commonality in concept allows for the cross-disciplinary work required to achieve community resilience, as discussed in the following.

While advancing resilience results requires a common understanding of resilience concepts, an inordinate amount of effort has been spent on seeking agreement on a formal definition. A formal definition is important as a guiding light, but, as concluded in the first chapter, a high-level definition will not lead to operational resilience, but agreement on the two required themes of minimizing impact and reducing time to recovery, common to most definitions, will lead there. Efforts should focus on developing the codes, standards, metrics, benchmarks, and other guidance that drive achievement of resilience. Development of such guidance will require a multidisciplinary effort. This chapter captures information on some of the activities underway in an attempt to inform practitioners and encourage them to provide expertise as such efforts evolve.

Recognizing the need for a coordinated and collaborative approach to resilience, the American Institute of Architects (AIA) and NIBS initiated an effort to bring together participants representing all aspects of the planning, design, construction, regulation, operation, and management of buildings. The effort resulted in the development of the *Industry Statement on Resilience* initially signed by almost two dozen organizations in 2014 and in 2019 reached almost 50 organizations (AIA 2019). ASCE and the International Code Council were among the initial signatories.

To support general understanding and agreement, the *Statement* points to the definition offered by the National Academies in its 2012 publication *Disaster Resilience: A National Imperative* (NRC 2012). The Academies defined resilience as "The ability to prepare and plan for, absorb, recover from, or more successfully adapt to actual or potential adverse events." As the Academies point out, this definition is consistent with the definition used in the international disaster policy community through the United Nations International Strategy for Disaster Reduction (UNISDR) and US governmental agencies including through Presidential Policy Directive 8 on national preparedness (UNISDR 2011, White House 2011).

In addition to the written definition offered by the Academies, the Alliance for National & Community Resilience (ANCR) has developed a graphical translation that helps convey the concept. This approach is featured in Figure 3-1. In its steady-state condition, the community operates on a trajectory. This trajectory represents the level of functionality as a function of time. In some cases, the community is in decline or under stress, in others it is improving. A shock (whether a disaster event or a

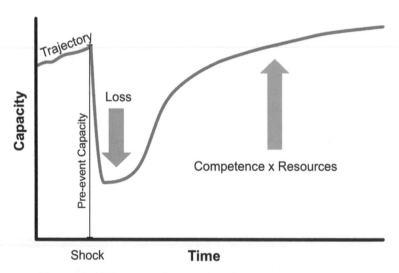

Adapted from UK Department for International Development

Figure 3-1. Graphical representation of resilience.
Source: Plodinec (2019).

social disruption) can significantly alter a community, triggering a need to recover. The significance of the disruption and the time needed to return to "normal" (or even a better state) represents the community's resilience.

For the sake of assuring a common understanding, this chapter will utilize the Academies' definition for resilience but recognizes that most other definitions containing similar concepts are equally valid and sufficient for high-level discussion. The development of actionable guidance and associated metrics will ultimately lead to the achievement of resilience. This chapter works to peel back the layers to get to the achievement of objective resilience.

3.1.2 Scale

Communities develop and function at a variety of scales. Designers and their clients often focus at the individual project level (unless engaged in a master planning effort). Incorporating resilience measures is important at both the project scale and the community scale, but when done in isolation, the end result may not be optimal. Resilience must be considered across scales and with engagement of the designers and decision makers that influence choices at each level. Practitioners should recognize the "bigger picture" and how their project-level decisions influence the resilience of the community and how the community can impact the performance of their project (Anderson 2019).

ANCR recognized that current resilience efforts are largely focused within individual disciplines or systems. Yet, community resilience requires a holistic approach, one that does not overinvest in the resilience of one sector while underinvesting in another. A community is only as resilient as its weakest link. ANCR identified 19 community functions that make communities what they are and, if they fail, reduce the resilience of the community. These functions cut across social, organizational, and infrastructural aspects of communities (Figure 3-2).

To support a community's holistic examination of resilience, ANCR is developing metrics and benchmarks in each of the functional areas. The metrics and benchmarks, to the greatest extent practical, rely on codes and standards. This allows for consistency in determining achievement of results and supports a defensible process that can be incorporated into other processes, including insurance and financial underwriting, grant programs, and corporate investment decisions.

3.2 COMMUNITY RESILIENCE

The definition of resilience can be applied to multiple systems on multiple scales. This allows recognition that each community function has

Figure 3-2. Community functions identified by the ANCR.

its own resilience (often defined and achieved by disciplinary actors within those functions) and that the system as a whole also has its own level of resilience (currently largely defined subjectively by policy makers). Achieving resilience at a community level requires examination and coordination across all community functions. Codes and standards are an example of a mechanism that supports a common, objective starting point for a coordinated approach.

3.2.1 Codes and Standards

Since the time of Hammurabi around 1800 BCE, the performance of buildings has been regulated through the application of codes. The Code of Hammurabi provided a set of performance requirements for buildings along with remedies for noncompliance (Avalon Project 2008):

- If a builder build[s] a house for someone and complete[s] it, he shall give him a fee of two shekels in money for each sar (a sar is an ancient unit of measurement equal to an area 12 cu. sq. or approximately 30 sq. ft) of surface.
- If a builder build[s] a house for someone, and does not construct it properly, and the house that he built fall[s] in and kill[s] its owner, then that builder shall be put to death.

- If it kill[s] the son of the owner, then the son of that builder shall be put to death.
- If it kill[s] a slave of the owner, then he shall pay slave for slave to the owner of the house.
- If it ruin[s] goods, he shall make compensation for all that has been ruined, and inasmuch as he did not construct this house properly that he built and it fell, he shall re-erect the house from his own means.
- If a builder build[s] a house for someone, even though he has not yet completed it, if then the walls seem toppling, the builder must make the walls solid from his own means.

Modern building codes were born out of disastrous fires that impacted major cities, including Boston (1631), London (1666), Chicago (1871), Baltimore (1904), and Cleveland (1929) (Vaughan and Turner 2013). Expanding from their initial roots in preventing the spread of fire, building codes now cover a suite of desired performance levels generally captured under the concepts of health, safety, and welfare (HSW). Today's building codes address both chronic and acute hazards. Provisions address acute risks posed by earthquakes, tsunamis, wildfires, hurricanes, floods, tornadoes, landslides, and blizzards and chronic risks such as increasing temperatures, termite infestation, and occupant health impacts. Additional HSW concepts covered in the codes include energy and water efficiency and indoor environmental quality (Davis and Ryan 2019).

The most widely used building codes in the United States are model codes produced by the International Code Council (Code Council). The suite of International Codes (I-Codes) includes 15 codes covering general requirements for commercial and residential buildings (the International Building Code and the International Residential Code), fire protection (International Fire Code), energy efficiency (International Energy Conservation Code), plumbing (International Plumbing Code), mechanical systems (International Mechanical Code), and wildland fire (International Wildland Urban Interface Code). The National Fire Protection Association (NFPA) also develops codes including the National Electrical Code (NFPA 70), a Fire Code (NFPA 1), and a Life-Safety Code (NFPA 101).

The Code Council uses a governmental consensus process that meets the principles of the *United States Standards Strategy* (ANSI 2015) and Office of Management and Budget (OMB 2016) *Circular A-119: Federal Participation in the Development and Use of Voluntary Consensus Standards and in Conformity Assessment Activities*. In addition, the governmental consensus process complies with the National Technology Transfer and Advancement Act of 1995 (NTTAA) (Public Law 104-113). Standards and model codes are a set of rules and guidelines designed to achieve a desired objective, usually safety. When a standard or model code is adopted into law by an authority having jurisdiction (AHJ), it becomes a code. Specifications are additional requirements of the owner regarding the materials, components that must be used in a building or other elements of infrastructure.

By their very nature, building codes provide an objective basis for determining the level of safety provided by the building covered. Building codes contribute to a baseline level of resilience at the building level and some elements of community resilience (more on this in the following). Codes are predicated on the ability to measure and evaluate the achievement of performance requirements. In many instances, the code sets the required level of performance and then points to standards that form the basis of objectivity including durability, fit for purpose, structural strength, impact resistance, and thermal and moisture conductance.

The I-Codes are supported by thousands of standards. These standards also, in general, meet the requirements contained in OMB Circular A-119 and the NTTAA. The expertise of organizations like ASCE is captured through the incorporation of standards such as *ASCE-7: Minimum Design Loads and Associated Criteria for Buildings and Other Structures* and *ASCE 24: Flood Resistant Design and Construction* (ASCE 2014, 2016).

While the I-Codes provide criteria for the design, construction, and operations of buildings, similar sets of requirements exist for other infrastructure, including bridges, water systems, pipelines, powerlines and substations, and railways. Many of these requirements are also often developed using the private sector–led standards development process. Organizations like the American Association of State Highway Transportation Officials (AASHTO), the American Water Works Association (AWWA), and the American Society of Mechanical Engineers (ASME) develop standards to support different infrastructure systems. While the content is different, most consensus-based standards rely on a common approach to verifying conformance.

Assuring that the structure and its components meet the requirements of the code relies on an enforcement and conformity assessment infrastructure. This infrastructure relies on the availability of replicable testing and evaluation. Products are tested, evaluated, and then labeled to indicate their ability to meet the provisions contained in the code. Designers, contractors, and code officials rely on these evaluations to help assure compliance.

Accreditation provides another layer of assurance. Where evaluation assesses whether a product conforms to a standard, accreditation assesses processes and practices—again against a specified standard or other criteria. Accreditation is available for multiple participants in the building process, from fabricators to inspectors and plan reviewers. Accreditation requires the development and implementation of quality assurance, management, and documentation processes. Once an entity has gone through the accreditation process, other participants in the building process have an objective measure of competence.

Internationally, codes and standards are often developed under the direction of a governmental standards body. While the desired outcomes are, in general, similar—protecting public safety—the process is different.

3.2.2 Continuum of Guidance

While codes are generally applied to individual buildings, they are a key part of a continuum of strategies that support resilient communities (Colker 2020). Without a recognition of this continuum planners, designers and community-level decision makers can perpetuate gaps in the achievement of community resilience. Designers often have an obligation through their professional codes of conduct to look beyond their project to understand its impact on the broader community and engage the project owner in discussions on how to mitigate such impacts.

The continuum of guidance for resilience runs from individual products and components to the structure, to the community, to the region, to the nation, and beyond (Colker 2020). The availability of coordinated and compatible objective measures all along this continuum is required to allow a rollup of measures to get the "complete picture" of resilience. Policy makers and practitioners need to assure that the policies and practices they deploy are aligned and aim to reach their objective resilience goals. Amorphous or subjective requirements leave results up to interpretation and potential disagreements on achievement (and the best means to get there).

Coordinating building codes with floodplain management policies (FEMA 2019), transportation plans, economic development plans, sustainability plans, hazard mitigation plans, and other community-level planning efforts is essential to assure the efficient realization of community resilience goals. Alignment of individual projects to the communitywide goals requires communicating these goals to those undertaking individual projects. Again, objectivity is essential to assure goals are communicated effectively and final results can be easily aligned with the goals.

Building codes are intended to translate communitywide goals for the performance of buildings down to the individual structure. Expectations for the safety of building occupants are inherently addressed, but the codes capture other benefits applicable to the community at large. Adjacent buildings are protected from impacts from neighboring buildings owing to collapse or flying debris in high winds; firefighters are assured access to prevent fire spread; plumbing systems are designed to prevent contamination of the community's water system; and sites are designed to prevent runoff into waterways. Other community policies and practices should pick up where the building codes leave off.

As discussed in the following, the coverage of buildings codes is evolving to meet changing needs including the expanding expectations of building performance. Leveraging the considerable expertise involved in the codes and standards development process to develop an expanded suite of building centric criteria will help advance achievement of community resilience. Designers and researchers have an obligation to engage in activities to improve the professions and the communities they serve.

3.2.3 Expanding Scope

To date, most codes and standards have focused on the immediate life-safety of building occupants—what measures should be taken to assure the building will not injure occupants in everyday use and will not cause harm in the event of a reasonably certain disaster event. While there is always room for additional improvements, the codes have made great strides in achieving these objectives. A FEMA analysis from 2014 estimated approximately $500 million in annualized loss avoided in eight southeastern states because of the adoption of modern building codes (FEMA 2014). A McClatchy analysis following the California Camp Fire in October 2018 found that 51% of the structures built after a wildland–urban interface fire code was implemented escaped damage compared with 18% of the 12,100 structures built prior (Kasler and Reese 2019). Postevent observations by FEMA, NIST, ASCE, and others frequently identify the vintage of structure (and, thus, the code edition it was built under) as a significant factor in the level of damage.

As interest in the concept of resilience grows, policy makers and building industry leaders have increasingly recognized that life safety is just one of multiple community-level goals. The resilience of a community following an adverse event relies on the social and economic viability of the community and not just the short-term ability for buildings to protect their occupants.

The recent ASCE Risk and Resilience Measurement Committee publication *Resilience-Based Performance: Next Generation Guidelines for Buildings and Lifeline Standards* clearly articulates the need to expand how codes and standards address the broader resilience needs of a community (ASCE 2019). "Current code-based standards are primarily focused on individual facilities and are out of sync with the resilience needs of the broader community. The emphasis on life safety and the lack of consideration of the consequences of loss of functionality will result in extensive socioeconomic disruptions and slow recovery after a major hazard event. A new generation of standards is needed, which redefines the current design approach such that it integrates community-level resilience goals with functional recovery-based design standards for individual facilities. It requires a convergent approach that brings together engineers, social scientists, economists, environmentalists and more."

The emerging concept of functional recovery is intended to capture this expanded recognition. Initial efforts to develop functional recovery standards have centered around seismic risk and been led by the state of California, the federal National Earthquake Hazard Reduction Program (NEHRP) through the National Institute of Standards and Technology (NIST) (NIST 2018, PUC 2019), and the Earthquake Engineering Research Institute (EERI) (EERI 2019). In July 2019, the Code Council and the California Building Officials (CALBO) hosted a workshop bringing together leaders in seismic codes and standards to identify a coordinated approach to the development of such criteria. While initially focused on earthquake hazards, functional recovery strategies and standards will develop across multiple hazards.

Implementing a broader focus beyond traditional life safety–based design will require a new mindset for both designers and regulators. Both must recognize that an individual structure possesses its own resilience but also contributes to the resilience of the community. Considerations at both scales are necessary. AIA has called for a redefinition of the concept of HSW that drives professional practice to be more holistic (AIA 2019). This holistic HSW definition would capture obligations associated with resilience, equitable design, and advanced building performance. Practitioners and the guidance, training, education, and regulation that support them must evolve to effectively address these changing needs.

3.2.4 Evolving Risks and Uncertainties

Codes and standards have evolved as a mechanism to address hazards and risks. As new knowledge is gained, codes and standards are updated to incorporate new solutions or new technologies. There is growing recognition within both the hazards community and the buildings community that the risks of the past are no longer the determinant of future risks. The usual customary assumption of stationarity for meteorological data (wind speeds, precipitation, temperature, and so on) is no longer valid in light of the changing climate (Reilly 2019).

Multiple factors are influencing future risks faced by society and the infrastructure they depend on. Addressing the resilience of infrastructure in the face of evolving climate-related risks presents a particular challenge.

Codes and standards developers and professional societies have identified multiple challenges in developing criteria to support infrastructure responsive to changes in risk associated with climate change:

- General lack of understanding of how to use climate data and models to inform infrastructure design.
- Lack of an authoritative source of future climate data and existing data is not in an actionable format.

- Limited precedents of others addressing these issues.
- Increasing recognition of the interdependence of infrastructure and associated planning points to the need for a common/compatible baseline across infrastructure systems.
- Historically limited interaction and engagement between the climate science and the building science community.
- Intended life of the structure impacts the potential exposure to future conditions and raises uncertainties as to the exact conditions it will face over its life cycle.
- Ad hoc approaches developed by individual practitioners may lead to liability exposure for individuals and create confusion for infrastructure owners and decision makers.

Policy makers and resilience practitioners at large need some organizing principles and strategies to lay the groundwork. Fortunately, there are some efforts underway to address these challenges. ASCE has taken the lead in examining how climate-related risk is changing both the design process and how codes and standards must evolve (Ayyub 2018, ASCE 2015). The Hoover Institute brought together subject matter experts in late 2018 to identify a potential path forward. Through that effort, they identified four principles and seven strategies for climate resilient infrastructure that were ultimately endorsed by organizations including ANCR, AIA, ASCE, and National Council for Science and the Environment (Hoover 2019).

The four principles to guide the development of more resilient infrastructure are as follows:

1. Be proactive. Do what we can now with both existing knowledge and foresight. Uncertainty should not preclude action.
2. Be fair. Consider the implications of decisions for those who are particularly vulnerable. We need to directly and consistently engage affected communities in decision-making.
3. Be inclusive. Engage all stakeholders early and often throughout the entire process. They should include knowledge generators, knowledge users, and impacted communities. An inclusive process helps to ensure that decisions are grounded in the best available information and fit the needs and values of those affected. Inclusivity can also reduce future conflict, avoid negative unintended consequences, identify a strong pool of options, and increase support for the measures chosen.
4. Be comprehensive. Consider the full range of risks and means to address them through planning, financing, and engineering. A holistic approach includes integrating social and ecological resilience into decisions where appropriate. Strong social dynamics and healthy, functioning ecosystems are critical to adaptive capacity—increasing

communities' and regions' ability to respond effectively to both chronic stresses and extreme events.

The seven strategies for climate-resilient infrastructure are as follows:

1. Make better decisions in the face of uncertainty.
2. View infrastructure systematically.
3. Take an iterative, multihazard approach.
4. Improve and inform cost–benefit analysis.
5. Mainstream nature-based infrastructure.
6. Jump-start resilience with immediate actions.
7. Plan now to build back better.

Engineers and other designers have unique experience and expertise to offer in the evolution of codes, standards, and practice to address these evolving needs. They can alert clients to shifts in risk and the need to protect assets across their life cycle, recognizing the uncertainties on climate impacts at the project scale. They should identify opportunities to build in resilience to the reasonably foreseeable risks and build in the opportunity to adapt to a reasonable suite of future conditions.

As codes, standards, and design practice evolve to address these risks, subjectivity will fade, leading to an enhanced understanding and implementation of new norms of professional practice to address evolving risks.

3.2.5 Metrics and Benchmarking

Resilience is an ever-progressing goal. Particularly with the changing risk associated with climate change, new challenges emerge. New research leads to improvements in the cost-effectiveness of various resilience strategies or the identification of new strategies all together. Research also reveals new risks. Further, as society addresses the most likely of risks, resources may become available to address less frequent or more diffuse impacts. Metrics and benchmarking are essential for planning, measuring progress, evaluating investment decisions, and reporting results. Metrics and benchmarks should exist at multiple scales but should be designed to allow understanding across scales and support the ability to "roll-up" metrics from multiple scales to get a higher level picture of progress. In addition, to the extent practical, metrics should be relatable across systems—building-level metrics should be compatible or relatable to metrics for energy systems for instance.

As identified by Brashear et al. (2018), the selection of the type of metric is essential to support compatibility and decision-making. For technical-level analysis, ratio scales provide a mechanism for comparison and integration across evaluation processes. Ordinal and nominal scales are best used to translate technical findings to the public and policy makers. Ratio scales have equal intervals (e.g., the distance between 1 and 2 is the same as that between 75 and 76) and a true zero point, meaning the absence of the quantity, thus enabling all mathematical functions. Ordinal scales (directional magnitudes, but not necessarily equal intervals, e.g., rankings, preferences on a five-point scale), and nominal scales (differentiating, but not ordering) have more limited application of mathematical functions, so they have limited utility in risk analysis and can contribute to distorting decisions. Ratio scales of measurement permit the full range of mathematical functions (e.g., can be added together or divided legitimately) and are clear in their meaning across users, systems, and organizations, for example, in information sharing for interdependency analysis.

The NIBS study presented here provides one valuable mechanism to evaluate decisions that lead to resilience—largely based on the monetary costs and benefits. The benefit–cost ratios (BCRs) provided represent a rigorous analysis based on significant peer review. The methodology is publicly available and can be applied to multiple types of mitigation strategies. The BCRs capture areas of benefits and costs where a significant body of academic study exists. These areas include property loss and damage; direct and indirect business disruption; injuries, deaths, and cases of post-traumatic stress disorder (PTSD); insurance administrative costs; and search and rescue costs. Naturally, additional costs are borne by individuals and society that cannot be effectively captured in such an analysis—the loss in educational achievement owing to disruptions in school attendance or owing to PTSD; the loss of cultural and historic resources; the loss of environmental services; and the disproportionate impacts that are felt by vulnerable populations. These additional factors reveal that the current BCR numbers are conservative and that additional factors can be considered in resilience investment decision-making.

In many cases at both the project and the policy levels, decisions are driven by cost, so the BCR results from the NIBS study are a reasonable factor to consider but should not be the only factor. Benefits and costs may accrue to different parties in different time frames, so additional analysis and evaluation (such as multicriteria decision analysis) may be useful. Community resilience relies on the social, organizational, and infrastructural resilience of community functions individually and as a whole. In most cases, determining social and organizational resilience requires a focus on factors that are difficult to evaluate on their own. Metrics associated with the achievement of codes and standards can be an effective proxy.

To the greatest extent practical, ANCR is relying on existing codes and standards as the basis for determining resilience. Its first benchmark covers buildings, recognizing that buildings support the functionality of many other community functions. An educational system relies on schools. Health care is provided inside hospitals and clinics. Emergency response functions are housed in police and fire departments. The Buildings Benchmark places significant emphasis on building codes and other standards to determine the resilience of a community's building sector (ANCR 2019a). ANCR's Housing Benchmark supports the social resilience provided by the availability and affordability of housing and shelters, again looking to established standards (ANCR 2019b). The Water Benchmark provides communities with important criteria for the resilience of drinking water, wastewater, and stormwater systems (ANCR 2020).

3.3 RECOMMENDATIONS

Engineers and other designers are on the front lines of implementing projects that either contribute to or detract from a community's resilience. They engage building and infrastructure owners and managers in a trusted relationship. They also typically have public-focused obligations placed on them as a function of licensure. Given the challenges before society to enhance resilience—particularly in the face of growing risks—engineers as a profession and individually have an opportunity to shape the solutions. The following recommendations set a path forward.

- Engage clients in meaningful discussion on the desired level of resilience for each project including an examination of codes and standards in place within the local jurisdiction; the continuum of guidance available including up-to-date codes, overlay codes, standards, and rating systems; the intended life cycle of the project; the potential for evolving risk over the project life cycle and opportunities to build in adaptability or enhancements to address this risk; and the project's context within the broader community including the impact it may have on community resilience and how it is impacted by the community's resilience. Such discussions shall inform the project design.
- Wherever they exist, designers should utilize codes, standards, metrics, benchmarks, and other guidance that support objective determination and expression of resilience at multiple scales from materials selection to an individual project to the whole community.
- Participate in efforts to enhance the profession and the communities you serve through engagement in codes and standards development and adoption; participating in community planning or policy development activities; educating the public on resilience; mentoring new or potential entrants into the building industry; and supporting research activities within your projects.

3.4 CONCLUSION

Effective and efficient achievement of community resilience requires approaches and processes based on science that allow consistency and objectivity. Codes and standards paired with metrics and benchmarks and a robust conformity assessment infrastructure provide the consistency and objectivity desired. Assuring their widespread use allows practitioners, policy makers, insurers, financiers, and others a common basis to evaluate a project's resilience contributions to community-, regional- and national-level goals.

Engineers and other designers play an essential role in communicating risk and options to address these risks to their clients. They also serve at the intersection of society and their clients. Designers should leverage these relationships to support effective and efficient achievement of resilience goals.

Objective resilience relies on the robustness of codes, standards, metrics, benchmarks, and conformity assessment.

BIBLIOGRAPHY

The American Institute of Architects has developed resources for their members to better understand resilience concepts and how architects in particular can incorporate resilience measures into their projects.

- AIA (American Institute of Architects). 2017. *Disaster assistance handbook*. Washington, DC: AIA.
- AIA. 2018. *Designing for resilience*. Washington, DC: AIA.

ASCE has adopted a policy statement outlining key definitions for critical infrastructure and resilience. This forms a useful baseline for discussion within ASCE, but an update is probably needed given the current state of resilience activities.

- ASCE. 2013. *Policy Statement 518—Unified definitions for critical infrastructure resilience*. New York: ASCE.

ASCE through the infrastructure report card and supporting document has made the case for increased investments in infrastructure. Content on the economic consequences of a failure to act can be particularly compelling to encourage action.

- ASCE. 2016. *Failure to act: Closing the infrastructure investment gap for America's economic future*. New York: ASCE.
- ASCE. 2017. *2017 Infrastructure report card: A comprehensive assessment of America's infrastructure*. New York: ASCE.

The Environmental Security Technology Certification Program (ESTCP) is a research program of the US Department of Defense. It has identified the need to incorporate climate risk into infrastructure decision-making. It held a workshop to identify a path forward. The resulting white paper outlines a potential approach.

- ESTCP (Environmental Security Technology Certification Program). 2017. *Workshop report: Nonstationary weather patterns and extreme events: Informing design and planning for long-lived infrastructure.* ESTCP Project RC-201591. Alexandria, VA: ESTCP.

Additional studies on the benefits of codes:

- IBHS (Insurance Institute for Business and Home Safety). 2004. *Hurricane Charley: Nature's force vs. structural strength.* Richburg, SC: IBHS.
- Louisiana State University Hurricane Center. 2003. *Residential wind damage in Hurricane Katrina: Preliminary estimates and potential loss reduction through improved building codes and construction practices.*

Bond rating agencies have begun examining the intersection of resilience and finance.

- MIS (Moody's Investors Service). 2017. *Environmental risks: Evaluating the impact of climate change on US state and local issuers.* New York: MIS.
- S&P Global. 2017. *Aftermath: Assessing the impact on local government credit quality.* https://www.spglobal.com/our-insights/In-A-Storms-Aftermath-Assessing-The-Impact-On-Local-Government-Credit-Quality.html.

The Insurance Services Office (ISO) through its Building Code Effectiveness Grading Schedule (BCEGS) evaluates how well states and localities are doing to address building-related vulnerabilities through the adoption and enforcement of building codes. BCEGS is designed for use in setting insurance rates but does provide an objective metric of a community's implementation of building codes.

- Insurance Services Office. *National building code assessment report: Building code effectiveness grading scale.* 2019 ed. Jersey City, NJ: Insurance Services Office.

The National Institute of Building Sciences (NIBS) through its Multihazard Mitigation Council and Council on Finance, Insurance and Real Estate has been identifying opportunities to bring together various incentive streams, including insurance and finance, to develop a holistic strategy for mitigation investments.

- MMC/CFIRE (Multihazard Mitigation Council and the Council on Finance, Insurance and Real Estate). 2015. *Developing pre-disaster resilience based on public and private incentivization*. Washington, DC: National Institute of Building Sciences.

- MMC/CFIRE. 2016. *An addendum to the white paper for developing pre-disaster resilience based on public and private sector incentivization*. Washington, DC: National Institute of Building Sciences.

The US Global Change Research Program, supported by relevant federal agencies, conducts regular assessments of the state of knowledge on climate science and includes impacts on key segments of the economy, including cities and infrastructure.

- USGCRP (U.S. Global Change Research Program). 2018. *Fourth National Climate Assessment*. https://www.globalchange.gov/nca4.

The United Nations has undertaken initiatives to support resilience across member states. The sustainable development goals and the Sendai Framework provide a common starting point for action. It has also begun efforts to develop metrics to support resilience benchmarking.

- UN (United Nations). n.d. *Sustainable development goals*. New York: UN.
- UNDRR (United Nations Office for Disaster Risk Reduction). 2015. *Sendai framework for disaster reduction*. Geneva: UNDRR.
- UNISDR (United Nations International Strategy for Disaster Reduction). 2017. *Disaster resilience scorecard for cities: Preliminary level assessment*. Geneva: UNISDR.

The National Institute of Standards and Technology (NIST) undertook an effort to develop community resilience strategies to support city-level initiatives. The planning guide lays out a set of steps to begin such an effort.

- NIST (National Institute of Standards and Technology). 2015. *Community resilience planning guide for buildings and infrastructure systems volume II*. NIST Special Publication 1190. Gaithersburg, MD: NIST.

The National Oceanic and Atmospheric Administration (NOAA) has been tracking the number and cost of natural disasters causing $1 billion in damage.

- NCEI (National Centers for Environmental Information). 2018. *U.S. billion-dollar weather and climate disasters*. https://www.ncdc.noaa.gov/billions/.

The US Department of Homeland Security convened a workshop to better understand the role of design and construction in infrastructure resilience. The workshop identified 18 recommendations to enhance infrastructure resilience.

- DHS (US Department of Homeland Security). 2010. *Designing for a resilient America: A stakeholder summit on high performance resilient buildings and related infrastructure.* Washington, DC: DHS.

Effective resilience practitioners will need an understanding of how various infrastructure systems interact and knowledge on how to align systems in furtherance of community resilience. Thought leaders from various disciplines share their insights to support a holistic understanding.

- Colker, R. M., ed. 2019. *Optimizing community infrastructure: Resilience in the face of shocks and stresses.* Amsterdam, Netherlands: Elsevier.

REFERENCES

AIA (American Institute of Architects). 2019. *Disruption, evolution, change: AIA's vision for the future of building codes, design and construction.* Washington, DC: AIA.

ANCR (Alliance for National and Community Resilience). 2019a. *Buildings benchmark.* Washington, DC: ANCR.

ANCR. 2019b. *Housing benchmark.* Washington, DC: ANCR.

ANCR. 2020. *Water benchmark.* Washington, DC: ANCR.

Anderson, A. H. 2019. "The role of designers and other building practitioners in advancing resilience." In *Optimizing community infrastructure: Resilience in the face of shocks and stresses,* edited by R. M. Colker, 197–210. Amsterdam, Netherlands: Elsevier.

ANSI (American National Standards Institute). 2015. *United States standards strategy.* New York: ANSI.

ASCE. 2014. *Flood resistant design and construction.* ASCE/SEI 24-14. Reston, VA: ASCE.

ASCE. 2015. *Adapting infrastructure and civil engineering practice to a changing climate.* Reston, VA: ASCE.

ASCE. 2016. *Minimum design loads and associated criteria for buildings and other structures.* ASCE/SEI 7-16. Reston, VA: ASCE.

ASCE. 2019. *Resilience-based performance: Next generation guidelines for buildings and lifeline standards.* Risk and Resilience Measurement Committee. Reston, VA: ASCE.

Avalon Project. 2008. *Code of Hammurabi.* New Haven, CT: Yale Law School, Lillian Goldman Law Library.

Ayyub, B. M. 2018. *Climate-resilient infrastructure: A manual of practice on adaptive design and risk management.* Reston, VA: ASCE.

Brashear, J. P., P. Scalingi, and R. Colker. 2018. *A business process engineering approach to managing security and resilience of lifeline infrastructures.* Washington, DC: National Institute of Building Sciences. Submitted under Contract HSHQDC-14-00089 between NIBS and the U.S. Department of Homeland Security. 2015.

CLF (Conservation Law Foundation). 2018. *Climate adaptation and liability: A legal primer and workshop summary report.* Boston, MA: CLF.

Colker, R. M. 2020. "The ecosystem of resilience standards." *Stand. Eng.* 72 (3): 9–12.

Czajkowski, J., and K. Simmons. 2014. "Convective storm vulnerability: Quantifying the role of effective and well-enforced building codes in minimizing Missouri hail property damage." *Land Econ.* 90 (3): 482–508.

Davis, C., and J. T. Ryan. 2019. "Building codes: The foundation for resilient communities." In *Optimizing community infrastructure: Resilience in the face of shocks and stresses,* edited by R. M. Colker, 211–238. Amsterdam, Netherlands: Elsevier.

EERI (Earthquake Engineering Research Institute). 2019. *Functional recovery: A conceptual framework.* Oakland, CA: EERI.

FEMA. 2014. *Phase 3 national methodology and phase 2 regional study losses avoided as a result of adopting and enforcing hazard-resistant building codes.* Washington, DC: FEMA.

FEMA. 2018. *2018–2022 Strategic plan.* Washington, DC: FEMA.

FEMA. 2019. *Reducing flood losses through the international codes.* 5th ed. Washington, DC: FEMA.

Haimes, Y. Y. 2009. "On the definition of resilience in systems." *Risk Anal.* 29 (4): 498–501.

Holling, C. S. 1973. "Resilience and stability of ecological systems." *Annu. Rev. Ecol. Syst.* 4: 1–23.

Hoover Institution. 2019. *Ready for tomorrow: Seven strategies for climate-resilient infrastructure.* Stanford, CA: Stanford University.

ICC and ANCR (International Code Council and Alliance for National and Community Resilience). 2018. *Building community resilience through modern model building codes.* Washington, DC: ICC.

"Industry Statement on Resilience." Accessed December 27, 2018. http://aiad8.prod.acquia-sites.com/sites/default/files/2018-04/Resilience_Statement_2018-0410.pdf.

Kasler, D., and P. Reese. 2019. "'The weakest link': Why your house may burn while your neighbor's survives the next wildfire." *Sacramento Bee.* https://www.sacbee.com/news/california/fires/article227665284.html.

Meerow, S., J. P. Newell, and M. Stults. 2016. "Defining urban resilience: A review." *Landscape Urban Plann.* 147: 38–49.

MitFLG (Mitigation Framework Leadership Group). 2019. *National mitigation investment strategy.* Washington, DC: MitFLG.

National Technology Transfer and Advancement Act of 1995, P.L. 104-113. Accessed September 17, 2021. https://www.nist.gov/standardsgov/national-technology-transfer-and-advancement-act-1995.

NIBS (National Institute of Building Sciences). 2005. *Natural hazard mitigation saves: An independent study to assess the future savings from mitigation activities.* Washington, DC: NIBS.

NIBS. 2018. *Natural hazard mitigation saves: 2018 Interim report.* Washington, DC: NIBS.

NIST (National Institute of Standards and Technology). 2018. *Research needs to support immediate occupancy building performance objective following natural hazard events.* NIST Special Publication 1224. Gaithersburg, MD: NIST.

NRC (National Research Council). 2012. *Disaster resilience: A national imperative.* Washington, DC: National Academies Press.

OMB (Office of Management and Budget). 2016. *Federal participation in the development and use of voluntary consensus standards and in conformity assessment activities.* OMB Circular A-119. Washington, DC: OMB.

Plodinec, J. 2019. "Where are we? Why community-wide benchmarking is important." In *Optimizing community infrastructure: Resilience in the face of shocks and stresses,* edited by R. M. Colker, 239–246. Amsterdam, Netherlands: Elsevier.

PUC (Provisions Update Committee). 2019. *Resilience-based design and the NEHRP provisions.* FEMA IDIQ Contract HSFE60-15-D-0022. Washington, DC: National Institute of Building Sciences.

Reilly, A. C., and B. Ayyub. 2019. "Designing for resilient systems under emerging risks." In *Optimizing community infrastructure: Resilience in the face of shocks and stresses,* edited by R. M. Colker. Amsterdam, Netherlands: Elsevier.

Rose, A. 2004. "Defining and measuring economic resilience to disasters." *Disaster Prev. Manage.* 13 (4): 307–314.

Simmons, K. M., J. Czajkowski, and J. M. Done. 2018. "Economic effectiveness of implementing a statewide building code: The case of Florida." *Land Econ.* 155–174.

UNISDR (United Nations International Strategy for Disaster Reduction). 2011. *Terminology.* Geneva: UNISDR.

Vaughan, E., and J. Turner. 2013. *The value and impact of building codes.* Washington, DC: Environmental and Energy Study Institute.

The White House. 2011. "Presidential Policy Directive-8: National preparedness." Accessed December 27, 2018. https://www.dhs.gov/xlibrary/assets/presidential-policy-directive-8-national-preparedness.pdf.

CHAPTER 4

RESILIENCE MANAGEMENT OF EFFECTS OF HAZARD EVENTS

Milagros Nanita-Kennett, Eric Letvin, Zackary Kennett

4.1 OVERVIEW

The purpose of this chapter is to analyze the relationship between disaster mitigation and resilience as well as to show the economic costs and benefits of this effort to the federal government and society. The Federal Emergency Management Agency (FEMA) of the Department of Homeland Security (DHS) is the main federal government agency directly involved in disaster recovery, and as a result, it has a wider remit to influence and improve resilience throughout the nation. At the time of this chapter's publication, FEMA is undergoing a structural transformation because its involvement and response to disasters has increased exponentially during the last two decades. The agency is currently seeking the most effective ways to deliver disaster resilience mitigation programs to the homeland. This chapter provides an overview of the FEMA's structure and disaster response up to 2020, the year of this publication.

4.2 CREATION OF FEMA AND ENACTMENT OF THE STAFFORD ACT

Historically, the United States has been affected by many catastrophic events such as the 1900 Galveston hurricane, the 1906 San Francisco earthquake, and the 1928 Okeechobee hurricane. More recently, the nation has been affected by Hurricane Katrina (August 26, 2005), which has been categorized as the most deadly and destructive hurricanes in US history. This event was followed by Hurricane Sandy (October 29, 2012) and the sequential and catastrophic hurricanes that occurred in 2017, which include Harvey (August 25), Irma (September 19), and Maria (September 20).

Before the creation of FEMA, federal, state, and local governments were all required to respond individually to natural disasters. In many instances, more than 100 federal agencies (these include the Department of Commerce, the General Services Administration, the Nuclear Regulatory Commission, the Treasury Department, the Department of Housing and Urban Development, the DOD Defense Civil Preparedness Agency, and the US Army Corps of Engineers) were involved—in an uncoordinated manner—in the response of the same disaster event.

Natural Disaster Declarations

The Robert T. Stafford Disaster Act states that the President can make a disaster declaration right after being requested by a governor of an affected state. There are two types of declarations, both of which require Presidential authorization:

- Emergency declarations: An emergency declaration can be declared for any occasion or instance when the President determines federal assistance is needed. Emergency declarations supplement state and local efforts in providing emergency services, such as the protection of lives, property, public health, and safety, or to lessen or avert the threat of a catastrophe in any part of the United States.
- Major declaration: The President can declare a major disaster declaration for any natural event, including any hurricane, tornado, storm, high water, wind-driven water, tidal wave, tsunami, earthquake, volcanic eruption, landslide, mudslide, snowstorm, or drought, or, regardless of cause, fire, flood, or explosion, that the President believes has caused damage of such severity that it is beyond the combined capabilities of state and local governments to respond. A major disaster declaration provides a wide range of federal assistance programs for individuals and public infrastructure, including funds for both emergency and permanent work.

As early as 1962, the US Army Corps of Engineers determined that that there were about 5,000 flood-prone communities and that flooding (from strong rains, snow melting, storms, or runoffs) was the most pervasive natural disaster. The government reacted to these findings and Congress enacted the 1968 National Flood Insurance Act (42 U.S.C. §4001) that provided subsidized flood insurance to home and business owners. This public law was followed by the 1974 Disaster Relief Act (Public Law 93-288) of the National Flood Insurance Program (NFIP), which laid down the process of the President's disaster declarations, expanded insurance coverage, and established a federal system of financial assistance to state and local governments.

To reduce fragmentation and consolidate the federal disaster emergency preparedness and response into a single agency, Congress enacted 1978 Reorganization Plan Number 3 (3 CFR 1978, 5 U.S. Code 903). This law was activated in 1979 by Executive Order 12148, which assigned new responsibilities to FEMA and merged other missions provided by the then Defense Civil Preparedness Agency, the Federal Preparedness Agency, the Federal Insurance Administration, the Federal Disaster Assistance Administration, the National Fire Prevention and Control Administration, the National Fire Academy, and the Community Preparedness Program. The recently founded FEMA also absorbed several programs operating in the White House and others pertaining to earthquake, dam safety, and emergency warning.

Almost 10 years after the creation of FEMA, the Robert T. Stafford Disaster Relief and Emergency Assistance Act of 1988 (Public Law 100-707) was issued, becoming the main statutory framework for most federal disaster relief efforts (DHS 1988). The Stafford Act coordinates the federal assistance in conjunction with state and local and tribal governments, determines the federal assistance provided before and after a disaster, and identifies the eligibility to certain nonprofit organizations, see Figure 4-1.

It is essential to highlight that the terrorist attacks that occurred on September 11, 2001, deeply transformed FEMA. These attacks, considered the worst in the homeland and in the world, killed 2,606 people at the

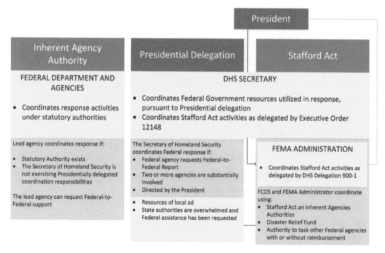

Note: FCOS refers to Federal Coordinating Officers who by Congress are responsible for managing disaster emergencies
Source: National Response Framework, May 2019

Figure 4-1. Current FEMA authority related to natural disasters.
Note: FCOS refers to Federal Coordinating Officers who by Congress are responsible for managing disaster emergencies.
Source: National Response Framework (FEMA 2019d).

World Trade Center area in New York City, 125 people at the Pentagon area in Northern Virginia, and 44 people on board of United Airlines Flight 93 which crashed in Pennsylvania after being highjacked with the intention to attack the White House in Washington, DC. In 2002, DHS was created by the Homeland Security Act (Public Law 107-296). It brought together 22 existing federal agencies, with FEMA among them.

4.3 KEY MITIGATION PROGRAM

Hazard mitigation measures are any sustainable action taken to reduce or eliminate long-term risk to people and property against disaster events. This section focuses on two major FEMA programs that provide funding for mitigation planning, disaster reduction losses, and protection of life and property from immediate or future damage: the Hazard Mitigation Assistance Program and the National Flood Insurance Program. These programs exemplify the extent of FEMA's national coverage in terms of resilience and the agency's efforts to break the cycle of disaster damage, reconstruction, and repeated losses, see Figure 4-2.

4.3.1 FEMA Hazard Mitigation Assistance

Federal funds are made available to state, local, tribal, and territorial governments in the form of grants. The HMA is comprised of three major grant programs: the Hazard Mitigation Grant Program, the Flood Mitigation Assistance Program, the Building Resilience Infrastructure and Communities and the Flood Mitigation Assistance Program.

- The Hazard Mitigation Grant Program (HMGP). The HMGP is a postdisaster grant authorized under Section 404 of the Stafford Act. The purpose of the HMGP is to help communities implement hazard mitigation measures following a Presidential major disaster

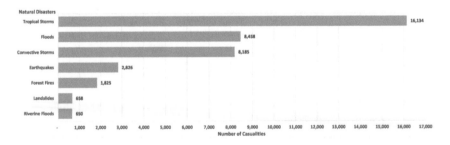

Figure 4-2. Natural disaster by number of casualties 1978–2017. Floods can be triggered by heavy rains, snow melting, storms, and urban runoffs. Flood and wind-related disasters are the most pervasive natural disasters in the United States.

Figure 4-3. Hazard Mitigation Grant Program by State 1998–2017. HMGP funding awards for recipients who have been impacted by large disasters.

declaration. The HMGP supports risk-reduction activities and aims at minimizing the impact of future disasters and providing cost-effective solutions during the period of disaster recovery. HMGP recipients have the primary responsibility for prioritizing, selecting, and administering state and local hazard mitigation projects. Although individuals may not apply directly to the state for assistance, local governments may sponsor an application on their behalf. After an application is made, it is evaluated, and if it meets eligibility requirements and funding is available, grant is awarded, see Figure 4-3. HMGP grant recipients will have 36 months from the close of the application period to complete projects.

Repetitive and Severe Flood Losses

Repetitive Flood Claims is a program authorized in 2004 to assist states and communities to reduce the potential for flood damage to properties that have had one or more claims to the NFIP. Only communities that are in a state with an approved state hazard mitigation plan are eligible to apply for and receive funds under this program. In June 2004, Congress passed the Flood Insurance Reform Act of 2004 (FIRA 2004), which contains provisions to develop a grant program for the mitigation of severe repetitive loss properties.

The HMGP was first implemented in 1989. Since then, till 2017, there were 21,638 awards for a total of $12.1 billion (FEMA data 10/2017). The State of Louisiana submitted the largest subaward totaling $729 million for Hurricane Katrina declared in 2005. The way that the HMGP is structured is that FEMA can fund up to 75% of the eligible cost of each project and 25% may come from nonfederal sources. Federal funding is based on 15% of the general funds spent on the public and individual assistance programs (FEMA programs that provide financial and direct services to eligible individuals and households affected by a disaster, who have uninsured or underinsured necessary expenses and serious needs) (minus administrative expenses) for each disaster declaration. Projects support state, tribal, and territorial priorities in their Hazard Mitigation Plan [This is a plan that is required as part of the FEMA grant award. The review guides of this plan are a part of the Stafford Act and the Code of Federal Regulations (44 CFR Part 201).] and may include voluntary acquisition of real property for conversion to open space and retrofitting measures such as elevation in place, structural relocation, and structural reinforcement.

- Building Resilient Infrastructure and Communities (BRIC). The BRIC program is a new FEMA predisaster hazard mitigation program that replaces the existing predisaster mitigation program. BRIC was authorized in 2020 by the Disaster Recovery Reform Act, Section 1234, amended Section 203 of the Stafford Act. The BRIC program aims to categorically shift the federal focus away from reactive disaster spending and toward research-supported, proactive investment in community resilience. FEMA anticipates BRIC funding projects that demonstrate innovative approaches to partnerships, such as shared funding mechanisms, and/or project design. The BRIC priorities are to

  - Incentivize public infrastructure projects.
  - Incentivize projects that mitigate risk to one or more lifelines.
  - Incentivize projects that incorporate nature-based solutions.
  - Incentivize adoption and enforcement of modern building codes.

BRIC funding levels will vary, as FEMA will calculate a 6% set aside within 180 days after each major disaster and set aside this amount from the Disaster Relief Fund for BRIC. The total amount will vary year to year based on the estimated amount of disaster assistance for each major Presidentially declared disaster and the number of Presidentially declared disasters in each year.

- The Flood Mitigation Assistance (FMA) Program. The FMA grant program provides resources to assist states, tribal governments, territories, and local communities in their efforts to reduce or eliminate the risk of flood damage to buildings and structures that are insured under the NFIP, see Figure 4-4. Eligible activities include property acquisition and structure demolition or relocation, structure elevation, dry flood-proofing, minor localized flood reduction projects, flood planning, nonstructural retrofitting of existing buildings and facilities, and the management costs of these activities, see Figure 4-5.

The PDM program was designed to provide fund to states, territories, Federally-recognized tribes, and local communities for hazard mitigation planning and the implementation of mitigation projects prior to a disaster event. The PDM has been used to fund non-structural mitigation projects such as hazard mitigation planning and removing structures from a flood plain that have repeatedly experienced flood damage. Some of the projects can include community safe rooms, drainage, and property elevation. The goal is to reduce the overall risk to communities and to break the cycle of disaster damage, reconstruction, and repeated damage.

The PDM program has experienced many changes in funding levels. It was created in 1997 by a Stafford Act amendment but it was not officially funded until 2009 when Congress introduced legislation (H.R. 1746). PDM grants were Congressional appropriations awarded on a nationally competitive basis. The PDM program was structured in such a manner that Federal assistance could contribute up to 75 percent of the total cost of mitigation activities and up to 90 percent in the case of small or impoverished communities. The total amount of funds distributed for PDM was determined once the appropriation was provided by Congress on a yearly basis. FEMA data shows that since 2001 3,029 PDM grants have been awarded at the cost of $1.39 billion. At present, this program has been subsetted.

Figure 4-4. The Pre-Disaster Mitigation (PDM) program.

FMA is authorized by Section 1366 of the National Flood Insurance Act of 1968, as amended (NFIA), 42 U.S.C. 4104c, with the goal of reducing or eliminating claims under the NFIP. FMA was created as part of the National Flood Insurance Reform Act (NFIRA) of 1994. The Biggert-Waters Flood Insurance Reform Act of 2012 (Public Law 112-141) (This was designed to allow premiums to rise to reflect the true risk of living in high-flood areas.) consolidated the Repetitive Flood Claims (RFC) [the RFC grant program provides funding to reduce or eliminate the long-term risk of flood damage to structures insured under the NFIP that have had one or more claim payment(s) for flood damage] and Severe Repetitive Loss (SRL) (the primary SRL properties' strategy is to eliminate or reduce the damage to residential property and the disruption to life caused by repeated flooding) grant programs into FMA. Approximately 9,000 insured properties have been identified with a high frequency of losses or a high value of claims.

Figure 4-5. The 1997–2017. Major recipients of the FMA grants are Texas, Louisiana, New Jersey, and Florida.

FMA grants are funded annually by Congressional appropriations from the National Flood Insurance Fund (NFIF) and are awarded on a nationally competitive basis.

About Private Flood Insurance and the NFIP

Private companies rarely provide flood insurance. This is because many private insurance companies stopped offering coverage for flooding following massive and expensive floods that plagued communities along the Mississippi River in the 1920s. In the decades that followed, Congress recognized the need to create a method for handling widespread flood damage when it occurs, passing flood insurance legislation in the 1950s and commissioning reports and studies in the 1960s. Since its creation, the NFIP is the nation's first line of defense against flood damage, with currently more than 5.3 million policies and more than $1.3 trillion in coverage in 56 states and participating jurisdictions.

The federal and nonfederal cost share depends on the type of properties included in the grant and determination in meeting the SRL and repetitive loss (RL) (An RL property is any insurable building for which two or more claims of more than $1,000 were paid by the NFIP within any rolling 10-year period since 1978. An RL property may or may not be currently insured by the NFIP.) definitions of 42 USC §4104c(h) per Biggert-Waters 2012. FEMA may contribute 100% cost share for SRL properties, 90% cost share for RL properties, and 75% cost share for properties that are NFIP-insured but do not meet SRL or RL definitions, as well as other project and planning types. The nonfederal share may be met with cash, contributions, and certain other grants such as Community Development Block Grants (HUD-CDBG) (The Community CDBG program provides annual grants on a formula basis to states, cities, and counties to develop viable urban communities by providing decent housing and a suitable living environment and by expanding economic opportunities, principally for low- and moderate-income persons.), Increased Cost of Compliance (The Increased Cost of Compliance coverage is one of several resources for flood insurance policyholders who need additional help rebuilding after a flood.) flood insurance payments, or in-kind services.

Flood Plain Management

FEMA is required to provide data on which flood plain management regulations of a particular community will be based. If these data are not available, the community may obtain and review reasonable data from other sources.

The symbols establishing special flood hazard designations are set forth in §64.3. If a proposed building site is in a flood-prone area, all new construction and substantial improvements will be designed (or modified) and adequately anchored to prevent flotation, collapse, or lateral movement of the structure resulting from hydrodynamic and hydrostatic loads, including the effects of buoyancy; be constructed with materials resistant to flood damage; be constructed by methods and practices that minimize flood damage; and be constructed with electrical, heating, ventilation, plumbing, and air conditioning equipment and other service facilities that are designed and/or located so as to prevent water from entering or accumulating within the components during conditions of flooding.

Policy holders may be required to purchase a separate insurance policy to offset any expenses of complying with more rigorous building codes [increase the Cost of Compliance (ICC)]. Congress has capped the amount that can be paid for ICC coverage at $75. The ICC policy has a separate rate premium structure and provides an amount up to $30,000 per policy.

States, territories, and federally recognized tribes are eligible to apply as applicants for FMA funds. Local governments are considered subapplicants and must apply to their applicant state, territory, or federally recognized tribe. Both applicants and subapplicants must have approved Hazard Mitigation Plans (FEMA requires state, tribal, territorial, and local governments to develop and adopt hazard mitigation plans as a condition for receiving certain types of nonemergency disaster assistance, including funding for mitigation projects.) and be in "Good Standing" with the NFIP to be eligible. Data show that from 1997 to 2017 (FEMA data collected by 10/2017), a total of 1,685 FMA grants have been awarded for a total of $90.6 million.

4.3.2 National Flood Insurance Program

The NFIP was created to reduce the impact of flooding by providing, at a reasonable rate, insurance to property owners, renters, and businesses and by encouraging communities to adopt and enforce floodplain management regulations. The NFIP was designed to transfer risk from the federal government to an insured party who is willing to take on the risk by paying a fee or a premium. Flood insurance is available to 23,000 communities where the agency enforces basic Flood Plain Management regulations (44 CFR, 59.1, 60.3, 60.4, and 60.5). Participation in the NFIP is based on an agreement among local communities stating that the federal government will make flood insurance available for new constructions to come up in FEMA's Special Flood Hazard Areas (SFHAs). The SFHAs are areas where the NFIP's floodplain management regulations must be enforced and mandatory purchase of flood insurance applies. Homes and businesses in high-risk flood areas with mortgages from government-backed lenders are required to have flood insurance. The FEMA National Investment Strategy reports that if you live in a 100 year floodplain, there is more than a 1 in 4 chance that you will be flooded during a 30 year mortgage. During a 30 year mortgage, you are 27 times more likely to experience a flood than have a fire. (From FEMA Mitigation Framework Leadership Group, August 2019.) The NFIP is the FEMA's largest and most complex mitigation program.

The NFIP is managed through its subcomponent, the Federal Insurance and Mitigation Administration (FIMA), which is responsible for managing a number of tools that facilitate the adoption and enforcement of local floodplain regulations, ordinances, risk-reduction mitigation measures, and the purchasing requirements for flood insurance in NFIP-participating communities.

The FEMA traditional risk mitigation management tools include the following (the FEMA Flood Map Service Center is an online location sponsored by the FEMA that facilitates to find all flood-hazard mapping products created under the NFIP. The service includes access to official NFIP maps and access to products and tools to better understanding flood risk):

- Flood Insurance Study (FIS) Reports: The reports are compilations and presentations of flood risk data for specific watercourses, lakes, and coastal flood hazard areas within a community. When a flood study is completed, the information serves to prepare maps and corresponding reports. The FIS Report contains detailed flood elevation data in flood profiles and data tables.

- Flood Insurance Rate Maps (FIRMs): These are maps produced by the FEMA, in partnership with the state and local governments, depicting flood hazard areas and risk premium zones. FIRMs include flood hazard areas that are within an area that has a 1% chance (In the 1960s, the US government decided to use the 1% annual exceedance probability flood as the basis for the NFIP.) of flooding in any given year. In addition, they identified areas furthered divided into insurance risk zones. FIRMs are used to set rates of insurance against risk of flood and to determine whether buildings are insurable against floods. These maps are also used in community planning, the insurance industry, and by individuals to gain knowledge on how a particular property is to be affected by flooding and what cost-effective measures can be put in place to reduce imminent risk.
- Flood Hazard Mapping Program. FEMA, in consultation with states, local governments, and appropriate federal agencies, has the responsibility to create detailed flood risk maps. These maps provide high-quality information, methods to better assess the risk from flooding, and tools to facilitate the adoption of mitigation actions by the community.

4.4 THE COST OF MITIGATION

The cost of mitigating natural hazards has increased over the years as explained in Figure 4-6. The United States has sustained 254 weather events between 1980 and 2018 where the overall damage/costs has reached or exceeded $1 billion, yielding a combined aggregate cost of more than $1.7 trillion (Smith 2019). The reasons behind these increases are numerous and are many times associated with climate change. They include increases in the frequency and intensity of natural hazards or/and expansion of the built environment, as well as increases in the population and infrastructure urban and coastal areas. Typically, the extent of the damage is determined by the severity of the event, the community's preparedness, the strength of the built environment, and the response time to recovering community functions. All these factors interact together to determine the community's resilience.

4.4.1 Disaster Relief Fund

The Presidential Disaster Declarations constitute the primary federal mechanism to trigger funds to the DRF as established in the Stafford Act as amended (42 U.S.C. 5121 et seq.). The DRF is the primary source of

An important part of the Flood Plain Management are the flood zones that FEMA is required to defined according to data related to varying levels of flood risk. These zones are depicted on FIRM or Flood Hazard Boundary Map. Each zone reflects the severity or type of flooding in the area. These designations cover from low and moderate risk areas to high risk areas exposed to coastal and riveting flooding. The ability to determine the risk of flooding is paramount in flood prevention. Scientist and engineers use statistical probability (chance) to put a context to floods. To determine these probabilities all the annual peak streamflow values measured at a stream gage are examined. A stream gage is a location on a river where the height of the water and the quantity of flow (streamflow) are recorded. The 1-percent annual exceedance probability (AEP) means that flood has a 1 in 100 chance of being equaled or exceeded in any 1 year. The 0.2-percent AEP means a flood of that size or greater has a 0.2-percent chance (or 1 in 500 chance) of occurring in a given year. Larger floods, such as the 0.2-percent, as tolerance for risk is reduced and increased protection from flooding is desired. The National Mitigation Investment Strategy reports only 32 percent of disaster-prone jurisdictions have adopted disaster-resistant mitigation strategies. The flood-hazard designated zones include the following:

Zone C, Zone X: Areas determined to be outside 500-year floodplain determined to be outside the 1% and 0.2% annual chance floodplains.

Zone B, Zone X500: Areas of 500-year flood; areas of 100-year flood with average depths of less than 1 foot or with drainage areas less than 1 square mile; and areas protected by levee from 100-year flood. An area inundated by 0.2% annual chance flooding.

Zone A: An area inundated by 1% annual chance flooding, for which no Base Flood Elevations (BFE) shown on the FIRMS and flood profiles have been determined.

Zone AE: An area inundated by 1% annual chance flooding, for which BFEs have been determined.

Zone AH: An area inundated by 1% annual chance flooding (usually an area of ponding), for which BFEs have been determined; flood depths range from 1 to 3 feet.

Zone AO: An area inundated by 1% annual chance flooding (usually sheet flow on sloping terrain), for which average depths have been determined; flood depths range from 1 to 3 feet.

Zone AR: An area inundated by flooding, for which BFEs or average depths have been determined. This is an area that was previously, and will again, be protected from the 1% annual chance flood by a Federal flood protection system whose restoration is Federally funded and underway.

Zone A1-A30: An area inundated by 1% annual chance flooding, for which BFEs have been determined.

Zone VE: An area inundated by 1% annual chance flooding with velocity hazard (wave action); BFEs have been determined.

Zone V(1-30): Costal flood with velocity hazard (wave action); BFEs have not been determined.

Note1: FEMA National Mitigation Investment Strategy Mitigation Framework Leadership Group, August 2019

Note 2: More information on FEMA Flood Plain Designated Areas is provided at: htpps://www.fema.gov/flood

Note 3: BFEs refers to regulatory requirement for the elevation or floodproofing of structures. The relationship between the BFE and a structure's elevation determines the flood insurance premium.

Figure 4-6. FEMA key flood plain designated areas.

funding for the federal government's domestic general disaster relief programs. This fund, which is managed by FEMA, is an appropriation account that is renewed with new funding annually. When the DRF is almost exhausted, Congress provides supplemental appropriations to replenish the account.

Having enough availability of funds in the DRF for disaster relief has been challenging. By 2004, the Congressional Research Service had reported that even in years with relatively few major disasters, it was not uncommon for the federal government to annually appropriate between $2 billion and $6 billion to help pay for new or ongoing recovery projects. Consistently with these increases, between 2000 and 2013, the annual appropriations rose significantly, creating the largest peak in history to respond to hurricanes Katrina, Rita, and Wilma that hit the southeastern part of the United States. In 2017, an important peak also occurred as appropriations were approved to respond to the 2017 events that included Hurricane Harvey, Hurricane Maria, and the California wildfires (CRS 2020a). By 2020, FEMA had planned to obligate $12.3 billion in DRF funding for ongoing recovery from past catastrophic disasters, including more than $6.2 billion for the four 2017 events, $513 million for Hurricane Sandy (FEMA 2013), and $200 million for costs from Hurricanes Katrina, Rita, and Wilma (2005). (CRS 2020a) (For more, see FEMA recovery.fema.gov.)

4.4.2 Amendments to the Stafford Act

Since the Stafford Act was first enacted, it has been amended several times, usually following a major natural disaster. Katrina created a large devastation along the Gulf Coast of Louisiana, Mississippi, and Alabama. At least 1,245 people died during the hurricane and subsequent floods. The total property damage has been estimated at $108 billion. Post-Katrina legislation to the Stafford Act expands the President's disaster assistance authority and vests FEMA with a larger authority to direct and coordinate a federal disaster response; to support precautionary evacuations and recovery efforts; to provide transportation assistance for relocating and returning individuals displaced from their residences; to increase the authorized percentage of federal contributions under the HMGP; to eliminate the statutory ceilings on financial aid for housing repair and replacement; to establish new flexibilities and policies for housing assistance and temporary housing; to expand assistance for repair of owner-occupied private residences and utilities; to expand the authorization for professional counseling and services for the reunification of families separated after an emergency or major disaster; and to establish provisions to make available competent interpretation and translation services, see Figure 4-7.

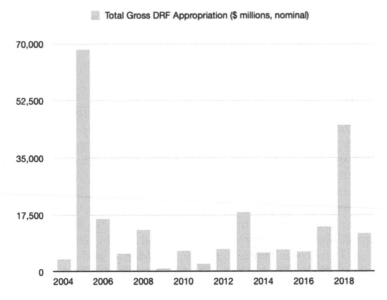

Figure 4-7. 2006–2019 Disaster relief appropriations.
Source: CRS (2019).

Similarly, the Stafford Act was amended after Superstorm Sandy that struck in October 2012. More than a dozen states were impacted by Hurricane Sandy. The storm affected heavily the Eastern Seaboard causing severe damage to New York and New Jersey. Sandy caused $18.75 billion in insured property losses, excluding flood insurance claims (NOAA, Hurricane Costs, Fast Facts, 2020). The storm damaged or destroyed more than 305,000 housing units and 265,000 businesses, and 8.5 million customers were left without power. [Hurricane Sandy and Disaster Relief, NY 2nd District 2013 (FEMA 2013)]. The storm battered the East Coast, particularly the densely populated New York, New Jersey, and Connecticut regions, with heavy rain, strong winds, and record storm surges and with heavy snowfall in West Virginia and the Appalachian Mountains. On January 29, 2013, the Disaster Relief Appropriations Act, 2013, a $50.5 billion package of disaster assistance largely focused on responding to Hurricane Sandy, was enacted as P.L. 113-2 (Painter and Brown 2017). The Sandy Recovery Improvement Act of 2013 authorizes FEMA to approve public assistance projects for major disasters or emergencies; increases the agency's flexibility in providing disaster assistance; expands federal assistance to a states, tribal, local governments, and owners or operators of a private nonprofit facilities;

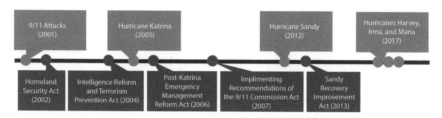

Figure 4-8. Timeline of major disasters.
Source: FEMA (2011).

and provides financial incentives and disincentives for the timely and cost-effective completion of projects.

The modifications made to the Stafford Act as a means to better respond to natural disasters and the large number of catastrophic events that have occurred since 2005 easily illustrate the reasons why the cost of responding to disaster has heightened consistently for more than a decade. Since being established in 1968 until 2004, the cost of the insurance program was mostly covered by the NFIP premiums. However, after Hurricane Katrina and later Hurricane Sandy, the NFIP has been forced to steadily borrow funds, accumulating by August 2017, $25 billion of debt. This situation was exacerbated by the 2017 hurricanes, Harvey, Irma, and Maria, which combined, produced more than quarter trillion dollars in damage in the United States. (National Hurricane Center reports in NOAA, 2017.) To respond to these disasters, Congress approved a disaster relief package that included a bailout of the financial difficulties of the NFIP, see Figure 4-8.

4.4.3 NFIP Losses

The desire to keep flood insurance affordable has presented many challenges for the NFIP. In January 2017, the NFIP borrowed $1.6 billion owing to losses (August 2016 Louisiana floods and Hurricane Matthew). On September 22, 2017, the NFIP borrowed the remaining $5.825 billion from the US Treasury to cover claims from hurricanes Harvey, Irma, and Maria reaching the NFIP's authorized borrowing limit of 30.425 billion. On October 26, 2017, Congress canceled $16 billion of the NFIP, but on November 9, 2017, FEMA borrowed another $6.1 billion to pay for losses incurred during hurricanes Harvey, Irma, and Maria, reaching again the authorized borrowing limit. US government records show the total loss estimates at $276.3 billion (CRS 2019).

At present and for more than a decade, FEMA has been facing the need to balance the competitive challenges of keeping flood insurance affordable to all participating communities, responding effectively to a large number of flood claims and keeping the NFIP fiscally healthy.

4.5 MITIGATION AND RESILIENCE

The term "mitigation" was firmly established by the enactment of the Stafford Disaster Act, which defines the term and creates many programs and grants. FEMA, which is the agency responsible for implementing most actions established in the Stafford Act, has traditionally used the word mitigation to define long-term actions to be taken before and after a disaster in an effort to save lives, reduce the impact of disaster events, and provide effective postdisaster reconstruction.

During the last decade, the term resilience has increased in usage within the disaster management community, lifting some of the central concepts of FEMA's mitigation programs. The term "resilience" is used to characterize actions that allow communities to bounce back and rapidly recover after a disaster event. Resilience can be defined as the ability of people and all elements of the physical environment to anticipate, absorb, and adapt to changing conditions and withstand and rapidly recover from negative impacts incompatible with fundamental human survival and needs. The effectiveness of a resilient environment promotes the capacity to reduce the magnitude and duration of disruptive events and the depletion of natural resources. Resilience core competencies can be broken down into four elements or 4Rs: robustness (R1), resourcefulness (R2), recovery (R3), and redundancy (R4).

Figure 4-9 presents an alignment of resilience's 4Rs with the traditional FEMA's core elements: preparedness, emergency response, mitigation, and recovery. This chapter argues that FEMA, through policies and programs, has been a key factor in paving the road for disaster management resilience and helping communities to "bounce back" and recover from hazardous and extreme events. Currently, FEMA has fully embraced the concept of resilience that is present in most FEMA programs.

4.6 FEMA'S NEW STRATEGY TOWARD RESILIENCE

In 2018, FEMA, in compliance with several federal directives and executive orders, launched a new Strategic Plan (2018–2022 FEMA Strategic Plan) that drives FEMA in the direction of forming a resilience framework, recognizing the fact that resilience is the backbone of emergency

Resilience Core Competencies	FEMA Core Mitigation Programs	
Robustness (R1): Inherent Strength and Resistance. It can be defined as the ability to absorb and adapt to changing circumstances and the ability to maintain critical operations and functions in the face of crisis.	**Mitigation.** It is an effort to introduce appropriate risk reduction measures, before and after a disaster strikes, in order to reduce the impact of disaster events to a physical environment and minimize the loss of life.	This includes the delivery of programs directed to mitigate the impact of natural hazards; to support the adoption of the latest model and/or local codes regulations, local ordinances, land use, building practices, and multi-hazard engineering science and advanced technology; to identify and minimize short- and long-term risks; and to adopt effective planning to manage the entire life cycle of a potential hazardous event.
Resourcefulness (R2): Capacity to Undertake Appropriate Action. It can be defined as the ability to skillfully prepare for, respond to and manage a crisis or disruption as it unfolds	**Preparedness.** It refers to all activities and measures necessary to prepare for or minimize the effects of a hazard upon the civilian population. **Emergency response.** It refers to efforts to respond to a situation that poses an immediate risk to health, life, property, or environment.	This includes the establishment of appropriate organizations and programs; recruitment and training of personnel and first responders; preparation of training exercises; preparation evacuation of individual and families at risk; facilitation of procurement and stockpiling of necessary materials and supplies; preparation of Federal, state, and local government plans; preparation of supporting agreements at a Federal, state, and local government level; provision of suitable warning systems; and construction or organization of shelters, shelter areas, and control centers.
Recovery (R3): Ability to Bounce Back. It can be defined as the ability to return to and/or reconstitute normal operations as quickly and efficiently as possible after a disruption	**Recovery.** It refers to efforts to identify, capture, and coordinate resources and capabilities to assist individuals, civic institutions, businesses, and governmental organizations to return to normal life after a disaster and protect against future hazards.	These efforts aim at making the most effective use of the Nation's resources to help communities to "bounce back" for destructive hazardous events. It is guided to share a common and integrated perspective across all Federal agencies, state and local government, nongovernmental partners, and stakeholders to improve the access to resources and foster effective coordination. Currently FEMA operates under the Presidential Policy Directive 8: National Preparedness, which emphasizes the actions directed at the full recovery of communities affected by disaster events.
Redundancy (R4): Ability to Function Despite the Disruption. It can be defined as additional or alternative systems, sub-systems, assets, or processes that maintain a degree of overall functionality in case of loss or failure of another system, sub-system, asset, or process	**Continuity of Operations.** It refers to the identification and prioritization of essential functions and the continuation of such functions during and after a disaster.	This aims at staffing and identifying and allocating resources to allow the Federal Government to perform all of its normal functions during and after a disaster event in order to sustain the most effective assistance to communities impacted by natural and man-made disasters. FEMA currently operates under the 2017 Federal Continuity Directive which was originally created in 2004 for government functions.

Figure 4-9. Interrelationship between mitigation and resilience.

management. FEMA has largely incorporated the concept of resilience in most of its major programs.

The rest of this chapter highlights a series of perspectives, frameworks, and programs that FEMA has put forward as it moves toward resilience as a reason for improving disaster emergency response and recovery. These include the FEMA National Response Frameworks, the FEMA Economic Sectors and Resilience, the New FEMA Grant Program, and the NFIP Trajectory and New Resilience Initiatives.

The new FEMA approach evolves around the following:

- Climate change continues to impact communities while disaster costs are expected to continue their increase because of rising natural hazard risks, decaying critical infrastructures, and economic vulnerability within many communities.
- Marginalized communities and a more diverse population will create pressure on authorities to incorporate specialized needs, expectations, and methods of communications into their plans, which may have an impact on the cost of natural disasters.
- Construction of new buildings and renovation of existing buildings is a life safety issue that is primarily the responsibility of state and local governments. The goal is to encourage the adoption and enforcement of the most up-to-date building codes and to educate the public and policymakers as to the importance of following best practices.
- Spectrum of viable threats will push authorities to detect, defend, and adopt new technologies to protect mitigation and resilience programs against cyber threats and terrorism for which the acquisition and maintenance of new technologies may be necessary.

4.6.1 2019 FEMA National Response Frameworks

On August 2018, FEMA published a revised National Response Framework (NRF) [Fourth Edition (Draft), May 28, 2019] establishing a process to be used as a road map for building resilience into critical infrastructure assets, which ensures that the agency can sustain, nationwide, its mission-essential functions in times of threats and disasters, as well as during normal operations. The framework builds on six main factors:

1. Stakeholder engagement is the departing point of the framework. This stresses the need to base all important phases of a disaster in a system or organizations in which all members or participants are seen as having an interest in its success of a particular mission. The appropriate stakeholders depend on expertise, geographic location, size of the geographical location, and real property and mobile assets portfolios.

2. Identify a critical mission using business process analysis (BPA) to identify mission essential systems, functions, and their associated critical infrastructure mission essential assets (MEAs).
3. Conduct a criticality assessment using business impact analysis (BIA) to determine how important, or critical, are the identified mission essential functions and assets.
4. Assess liabilities by analyzing the level of risk posed by potential hazards and threats to, and vulnerabilities of, the mission critical functions and assets.
5. Identify resilience gaps and determine resilience solutions that will ensure that MEAs are sufficiently resilient so that no loss of critical mission essential functions occurs beyond the maximum tolerable downtime during and after disaster events. In association with the Resilience Framework, a Resilience Readiness Planning Assessment guide was developed to score the level of resilience within each of the four focus areas that can be applied to our specific sites.
6. Integrate resilience readiness solutions that will close the gaps between the current state and a resilient state of MEAs to ensure continuous performance of critical mission essential functions as needed during times of hazard or threat disruption, as well as during normal operations.

The FEMA NRF provides foundational emergency management for how the nation responds to all types of incidents. It establishes the fundamental doctrine to support locally executed, state-managed, and federally supported disaster operations.

Problems and Disparities in the Application of Building Codes

Natural disaster events may have a devastating effect on the built environment, creating large losses for owners as well as federal, state, and local governments. The disparity in the application of building codes can be clearly seen in aerial photographs taken after the occurrence of a disaster event. Constructions complying with appropriate codes and standards typically display minor damage, whereas communities that are not built to the standards of codes show significant damage or collapse. The loose application of codes jeopardizes lives, livelihood, and economic recovery.

FEMA acknowledges that there are barriers to code development, implementation, and effective code enforcement. Misconceptions about the impact of building codes on housing affordability constitute an important issue.

In addition, there are unfounded criticisms of codes from certain groups and some state code council decisions to delay or weaken their codes. Pushback from builders and manufacturers who benefit financially or otherwise from weaker building codes have stunted the inclusion of some disaster-resistant provisions into the codes.

Despite FEMA's positive efforts to strengthen consensus model building codes and standards over time, the agency is ill-equipped to engage in a sustained way as building code adoption issues threaten the growth of resilience in communities across the nation. FEMA participated in the updating of the Mitigation Saves[1] study, following which there is a significant improvement in the ratios; the 4 to 1 ratio references money spent by federal agencies. The new study contains new material that shows how funds spend on mitigation can save even more than 4 to 1.

---

[1]https://www.fema.gov/media-library-data/1528732098546-c3116b4c12a0
167c31b46ba09d02edfa/FEMA_MitSaves-Factsheet_508.pdf

- Community Lifelines

The FEMA NRF focuses on the importance of sustaining and restoring essential community lifelines. A lifeline can be defined as a system that provides indispensable service that enables the continuous operation of critical business and government functions and is critical to human health and safety, or economic security. The FEMA NRF identified seven critical community lifelines: safety and security; food, water, and shelter; health and medical; energy (power and fuel); communications; transportation; and hazardous material.

Community lifelines are interdependent and vulnerable to cascading failures. A cascading effect can be defined as an unforeseen chain of events owing to a disruption affecting one or more systems. The community lifelines framework may provide the opportunity to the disaster management community to analyze the possibility that a cascading effect may have a negative impact on a system or multiple systems that can lead to total or partial failure of one or more systems. For example, communications and electric power systems rely on each other to function; severe damage to one will disrupt the other. Most lifelines also rely on complex supply chains. For instance, water and wastewater service depends on the resupply of a broad array of chemicals and—if power goes out—fuel for emergency generators. However, in a severe natural or human-caused incident, those supply chains themselves may be crippled.

Community lifelines can be used by all levels of government, the private sector, and other partners to facilitate operational coordination and drive

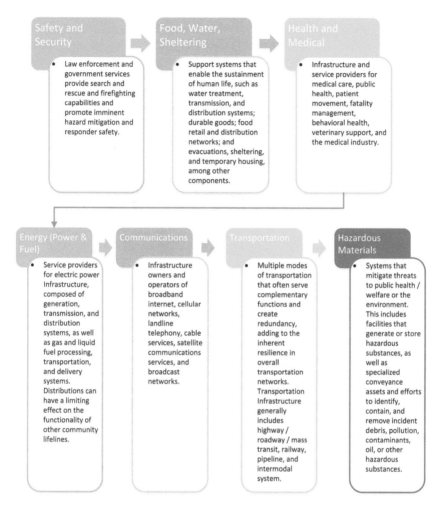

Figure 4-10. FEMA community lifelines.

outcome-based response. The scope is comprehensive and fits all the stages of natural disaster planning, emergency response, damage assessments, community support, and community stabilization. The application of the community lifelines is depicted in Figures 4-10 to 4-12.

4.6.2 FEMA Economic Sectors and Resilience

On June 2016, FEMA and the National Oceanic and Atmospheric Administration (NOAA) undertook a joint effort to explore resilience indicators and measures. The results of these efforts are included in a draft

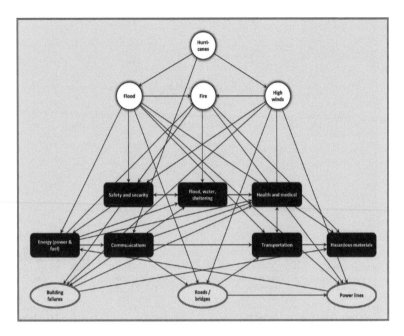

Figure 4-11. Example cascading effects FEMA community lifelines. This figure shows how a single hazard, hurricane, can trigger multiple effects on all seven lifelines, and then each of these seven lifelines will affect and interact with each other, either directly or indirectly by creating a cascading effect or cascading failure in different infrastructures (in this case, infrastructures: buildings/bridges/ roads/power lines).
Source: Dr. Mohammed Ettouney.

chapter entitled *Interagency Concept for Community Resilience Indicators and National Level Measures* (DHS 2016). The main objective of the methodology is to identify key attributes of community resilience (core capability) and the capacity related to the mitigation and recovery mission areas (indicators). The document illustrates potential options for measuring resilience.

Factors of the NOAA Economic and Resilience Taxonomy

- Geographic Scope—Indicator to be used to summarize community resilience capacity information and possibly track relevant trends over time at a national level. The indicator may also be used to develop measures at regional, state, and local scales.

- Geography—Indicators to reflect a wide range of community resilience factors that cut across different types and sizes of communities. The indicators represent resilience attributes in urban and suburban areas than in rural areas.
- Threat or Hazard—Indicators that focus primarily on community capacity to resist all hazards.

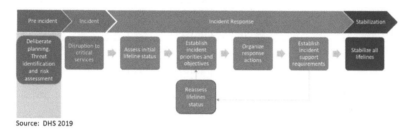

Source: DHS 2019

Figure 4-12. Application of community lifelines.
Source: DHS (2019).

Although unfinished, this methodology is a great framework to envision resilience as a system of systems where the core capabilities are applied to field sector indicators to understand their interactions to satisfy community requirements. The document includes several tables that include the measurement for possible uses of federal program–based information in identifying relevant national-level trends in community resilience capacity building activities and outcomes.

This framework can be very helpful to disaster management practitioners. The indicators can be used as a measure to assess resilience within a specific community. Crucial stages of measurement could include a baseline, postdisaster, and recovery measurement to quantify the impacts of that community's resilience. For instance, the condition of the housing stock can be used to see how the community normally operates with regard to occupied and available housing as well as the impacts on that by a natural disaster and how resilience plays a role in the changes of that quantifiable measure. The other indicators of this methodology are health and social services, economy recovery, infrastructure system, natural and cultural resources, threat and hazard identification, risk and disaster resilience assessment, planning, and community resilience, see Figure 4-13.

4.6.3 Modernizing the Delivery of FEMA Grants

FEMA currently manages more than 40 grant programs. As FEMA moves toward a more resilience framework, it is undertaking a wide

Core Capability		Indicators
1.	Housing	Housing Conditions
		Housing Affordability
2.	Health and Social Services	Health Care Availability
		Healthy Behaviors
		Environmental Health
3.	Economic Recovery	State and Local Government Revenues
		Employment Opportunity
		Income
4.	Infrastructure Systems	Roadway Conditions
		Transportation Connectivity
		Transit Accessibility
		Water Sector Emergency Support
		Energy Assurance
		Telecommunications Accessibility
		Dam Safety
		Integrated Infrastructure Sector Preparedness
5.	Natural and Cultural Resources	Water Conservation
		Wetlands Conservation
		Forest Conservation
		Habitat Quality
		Cultural Resources Protection
6.	Threat and Hazards Identification	Risk Identification
7.	Risk and Disaster Resilience Assessment	Risk Data
		Risk Awareness
		Community Preparedness
8.	Planning	Mitigation Planning
		Planning Integration
9.	Community Resilience	Collaborative Networks
		Civic Capacity
10.	Long-Term Vulnerability Reduction	Building Codes
		Higher Standards
		Mitigation Investment

Figure 4-13. FEMA/NOAA resilience taxonomy.

initiative to modernize and consolidate existing disparate FEMA grant management systems.

FEMA grant programs award billions of dollars every year before and after a disaster. Between the fiscal years 2008 and 2017, FEMA provided nearly $100 billion in financial assistance. In 2017, FEMA awarded more than $3.0 billion in preparedness and other nondisaster grants to support governments to prevent, protect against, mitigate, or respond to threats or incidents of terrorism and other events; more than $4.2 billion in individual assistance [Individual assistance is a program designed for individuals and families who have sustained losses because of disasters. Small Business Administration (SBA) disaster loans are the primary source of federal long-term disaster recovery funds for disaster damage not fully covered by insurance or other compensation.] including the individuals and households program, disaster case management, disaster legal services, disaster unemployment assistance, and crisis counseling program; more than $5.5 billion in public assistance (The Public Assistance Program provides supplemental grants to state, tribal, territorial, and local governments, and certain types of private nonprofits so that communities can quickly respond to and recover from major disasters or emergencies.) including funding to clear debris and rebuild roads, schools, libraries, and other public facilities; and more than $650 million in hazard mitigation grants (Hazard mitigation grants provide funding for eligible mitigation measures that reduce disaster losses.) to implement long-term hazard mitigation measures following a Presidential major disaster declaration. FEMA currently manages more than 40 grant programs to support DHS missions in response, recovery, mitigation, and resilience. The importance of the FEMA grant program is paramount.

- The Grant Modernization Program (GMM)

The GMM program is currently under design and its main purpose is to establish consistent and standardized business practices throughout the grant life cycle. This program aims to transform existing disparate FEMA grants management systems and business processes into a single IT platform, based on a streamlined and unified grants management life cycle using an iterative cloud development approach. The GMM program will provide the following core capabilities to support grants management across nondisaster, mitigation, and disaster operations: (a) business process management, (b) business rules management, (c) enterprise document and records management, (d) transaction management, (e) reporting and analytics, (f) grants case management, (g) collaboration support tools, and (h) hosting environment. See Figure 4-14 for an overview of GMM business goals.

The GMM allows for easier information sharing across programs to reduce collection requirements and eliminate redundant requests for data.

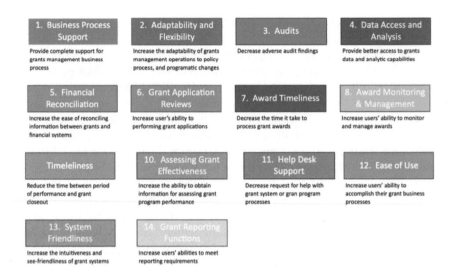

Figure 4-14. GMM business goals.
Source: FEMA (2020).

This program is expected to be operational in 2023 and aims at consolidating grants in four major buckets:

- Nondisaster competitive and formula grants;
- Predisaster mitigation grants;
- Postdisaster grants; and
- Individual assistance.

4.6.4 NFIP Trajectory and New Resilience Initiatives

The NFIP needs to be reauthorized by September 30, 2021, for its continuation (CRS 2020b). Reauthorization of the NFIP has occurred multiple times in the past. Since the end of 2017, 15 short-term reauthorizations have been enacted. By 2019, the US Treasury had authorized the NFIP to borrow up to $30.425 billion for its operations (CRS 2019). The underlying fact is that the NFIP has become more expensive than what was originally intended, accumulating significant losses every year. However, despite any fiscal problems in the past or at present, the reality is that the NFIP continues to be reauthorized by Congress as the program is an absolute necessity to maintain flood insurance that is affordable to all communities.

The Congressional Service Research reports that the issues that Congress may consider in the context of continuing reauthorizing the NFIP include the following: NFIP solvency and debt; premium rates and surcharges;

Implications of the Expiration of NFIP Authorities

The expiration of the NFIP's authority to provide new flood insurance contracts has potentially significant implications because of the mandatory purchase requirement (MPR). By law or regulation, federal agencies, federally regulated lending institutions, and government-sponsored enterprises must require certain property owners to purchase flood insurance as a condition of any mortgage that these entities make, guarantee, or purchase. Property owners, both residential and commercial, are required to purchase flood insurance if their property is identified as being in an SFHA (which is equivalent to having an estimated 1% or greater risk of flooding every year) and is in a community that participates in the NFIP. Without available flood insurance, real estate transactions in an SFHA potentially would be significantly hampered.

Source: Congressional Research Service.

---

affordability of flood insurance; increasing participation in the NFIP; the role of private insurance and barriers to private sector involvement; noninsurance functions of the NFIP such as floodplain mapping and flood mitigation; and future flood risks, including future catastrophic events.

In terms of the affordability of insurance policies offered by the NFIP, some argue that the NFIP is used as tax subsidies to unintentionally encourage building homes in dangerous locations. FEMA continues to work to correct current inequalities where sometimes wealthy coastal residents are paying rates that are lower than millions of other citizens located in less risky areas. Providing subsidized insurance policies in highly flood-prone and high-priced locations is perceived to be unfair to marginalized communities and impoverished individuals. This leads to financial losses because the NFIP does not benefit from selling certain policies at a higher rate. This type of policy creates incentives to expand and densify communities located in desirable coastal and riverine zip codes.

As FEMA advances toward resilience, it is faced with increasing unresolved challenges:

Servicing Policies and Claims Management

While FEMA provides overarching management and oversight of the NFIP, the bulk of the day-to-day-operations that include marketing, sales, writing, and claims management is carried out by private companies. There are two arrangements that FEMA has made with the private sector for which it is authorized by statue and guided regulation. The first one is the Direct Servicing Agent (DSA), which operates as a private contractor on behalf of FEMA for individuals seeking to purchase flood insurance policies directly from the NFIP. The second is the Write-Your-Own (WYO) program, where private insurance compares are paid to directly write and service the policies themselves. In either case, the NFIP retains the financial risk of paying claims for the policy and the policy terms and premiums are the same. Currently, approximately 13% of the total NFIP policy portfolio is managed through the DSA and 87% are sold by 59 companies participating in the WYO program.

Source: CRS (2019).

- Limited ability to reduce or eliminate damage to properties and disruption to life caused by the repeated flooding of SRL properties.
- Limited methods for identifying and recording when SRL properties have been mitigated.
- Limited ability to ensure timely and equitable distribution of mitigation funding to high-risk flood communities.
- Limited ability to provide uniform mitigation guidance, or oversight to enforce existing guidance.

- NFIP Pivot Program

FEMA is currently taking the approach to modernize several programs within the NFIP framework to increase efficiencies and maximize community resilience. Pivot is an FEMA flagship modernization program directed at replacing a more than 30 year-old system of records. Pivot automates a significant number of manual processes to improve efficiency and enable better information sharing and real-time analysis/reporting, such as policy sales, claims decision, and premium calculation. NFIP Pivot is designed to adopt robust data management standards and technology to enable enterprise-wide analytics. The Pivot system will provide immediate policy sale and claims decisions. The system will standardize and automate the risk-rating process (calculating cost to insure) so policyholders understand costs upfront. Pivot supports FEMA's role to help people understand risk and the available options to best manage risk. The primary Pivot user groups are internal groups within the Federal Insurance and Mitigation Administration (FIMA) and Write Your Own (WYO) insurance companies that sell flood insurance. See Figure 4-15 for an overview of the NFIP pivot target structure.

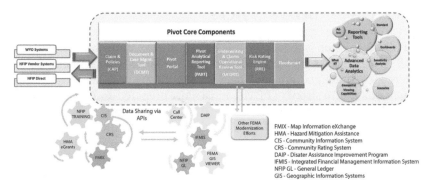

Source: FIMA

Figure 4-15. Pivot target structure.
Source: FEMA.

The program has developed a new state-of-the-art system that provides the capability to do the following:

- Develop innovative systems and business processes;
- Provide readily available and transparent customer service;
- Provide informing decision making by delivering capability to analyze NFIP data for response and recovery efforts, such as near real-time claims information and financial forecasting;
- Provide informing decision making by delivering capacity to analyze NFIP data for response and recovery efforts;
- Reduce manual reporting burden for companies during disaster response;
- Allow direct file uploads, reporting, and tracking capabilities;
- Provide insurance companies with capabilities to manage their own users, do self-care (e.g., password reset), and adjust permissions for their staff;
- Provide automated interfaces with other FEMA and government systems;
- Provide enhanced reporting and data analytics;
- Upload photos, elevation certificates, engineering reports, and other documents that are associated with a specific policy or claim;
- Securely process, transmit, store, and disseminate critical NFIP business information;
- Reduce the manual reporting burden;
- Serve as a single data store for all historical NFIP data; and
- Provide automated interfaces with other FEMA systems.

Pivot interfaces with many other FEMA systems to enable mission delivery, and the target architecture is included in Figures 4-10 to 4.12.

4.7 RECOMMENDED PRACTICES

FEMA's mission is to help people before, during, and after disasters. The 2018 to 2022 FEMA Strategic Plan creates a shared vision for the field of emergency management and sets an ambitious, yet achievable, path forward to build a stronger FEMA and a more prepared and resilient nation.

FEMA is committed to breaking the cycle of limited investment in resilience, "unprecedented" flooding, and repeated suffering by the same communities and populations. The devastating impacts on households, the cost of disaster assistance, and lost economic activity are all unsustainable. There is a need to shift from reliance on FEMA and the federal government to proactive action by communities and private sector leaders to mitigate risk and avoid future disaster-related losses (including floods).

Within the framework, this chapter recognizes the fact that the NFIP presents a unique opportunity to build a culture of mitigation and resilience to reduce the impact of disasters. The NFIP improves the nation's resilience to floods by

- Reducing the risk of injury and loss of life;
- Making homes and neighborhoods safer; and
- Strengthening lifelines (housing, energy, food, water, shelter, medical, safety, security, communications, transportation, and so on).

This chapter would like to make the following recommendations:

- FEMA should continue to build the capacity to make funding for mitigation investment easier to access. Both the GMM and the PIVOT are designed for this purpose (NIBS 2019). Efforts in this direction will mean that FEMA, in coordination with federal, state, and local governments as well as nonfederal partners, should simplify mitigation funding processes, coordinate cofunding, and encourage plan integration, all the while maintaining requirements and standards for effective mitigation and resilience measures.
- FEMA should create the capacity to consistently link mitigation investments and risk reduction by increasing the involvement of business owners, community officials, homeowners, and others.
- State-based and local-based authorities need to familiarize themselves with the new FEMA grants and flood insurance programs to minimize risk and take full advantage of these programs before and during disaster recovery.
- Private organization stakeholders (e.g., insurance companies, contractors, professional organizations, and so on) need to readjust their strategies to be consistent with new FEMA programs, regulations, and policies.

- FEMA should develop the capacity to link local risk by creating disincentives to build in high-risk coastal and riverine areas, and create incentive to build with appropriate codes and standards in safe areas using landscaping and environmental safeguards to reduce risk.
- FEMA should strengthen partnership and collaboration on a regional/national basis. Today, mitigation emphasis mostly encourages small-scale household or community actions. FEMA needs to bring federal, state, and private sector investors together to expand the use and impact of mitigation investments and resilience goals.
- Research and development by industry organizations, academic institutions, and consultants need to increase their familiarization with FEMA's new approach to mitigation and resilience. This may include areas such as management, social aspects, economic activities, engineering, architecture, security, and safety among many others. FEMA should increase its outreach capacity and communicate in clear and simple language with the population requiring flood insurance. This may include sharing a vocabulary that increases the whole community's understanding of mitigation methods and resilience goals. This may also include writing materials for people who are not frequently exposed to regulatory and technical guidance.
- FEMA should develop processes to encourage communities to adopt and enforce up-to-date building codes to regulate the design, construction, and occupancy of buildings and structures by providing minimum requirements to safeguard public safety, health, and general welfare. Architects, engineers, builders, and regulators should use the latest building codes (as prescribed by the relevant state and local government) for knowing about the most up-to-date requirements concerning structural integrity, mechanical integrity, fire prevention, and energy conservation. Using up-to-date building codes helps communities survive, remain resilient, and continue to provide essential services after a disaster occurs.
- Code development and enforcement vary widely across the country. Therefore, FEMA should strengthen its ability to keep building codes, specifications, and standards up to date because better codes improve the quality of buildings and make communities more resilient. It has been proven that the benefits of investing $1 in disaster mitigation provide a return of $11. (National Institute of Building Sciences Issues Interim Report on the Value of Mitigation, January 8, 2019.)
- FEMA should streamline the efforts that allow potential applicants to receive clear information for individuals as well as state and local governments requesting assistance before and after a disaster. The goal is to make sure that processes supporting funding are easy to understand, consistent across similar programs, and can

accommodate local applicants. FEMA with federal and nonfederal partners may encourage communities exposed to natural hazards to do the following:

- Develop resources and publications to inform the people and organizations that administer and enforce building codes.
- Support efforts that educate communities about the value of improving codes and practices while creating demand at the community and individual levels, such as the Federal Alliance for Safe Homes or the Institute for Business and Home Safety.
- Take actions that lead to updated state and local codes and practices.

BIBLIOGRAPHY

Bleemer, Z., and W. van der Klaauw. 2017. *Disaster (over-) insurance: The long-term financial and socioeconomic consequences of Hurricane Katrina.* Staff Rep. No. 807. New York: Federal Reserve Bank of New York.

Brown, J. T. 2011. *Presidential Policy Directive 8 and the national preparedness system: Background and issues for congress.* Washington, DC: Congressional Research Service.

CRS (Congressional Research Service). 2013. *Analysis of the Sandy Recovery Improvement Act of 2013.* Washington, DC: CRS.

CRS. 2015. *Mitigation assistance guidance, hazard mitigation grant program, pre-disaster mitigation program, and flood mitigation assistance program.* Washington, DC: CRS.

DHS (Department of Homeland Security). 2006. *Public Law 109-295.* Washington, DC: DHS.

DHS. 2013. *Public Law 113-2.* Washington, DC: DHS.

DHS. 2019. *National response framework.* Washington, DC: DHS.

Ettouney, M. M. 2014. *Resilience management: How it is becoming essential to civil infrastructure recovery.* New York: McGraw Hill.

FEMA. 2010. *Public assistance alternative procedures (Section 428). Guide for permanent work.* FEMA DR-4339-PR. Washington, DC: FEMA.

FEMA. 2011. *Enterprise common controls catalog data center 1 service level two.* Washington, DC: FEMA.

FEMA. 2012. *Adoption of flood insurance rate maps by participating communities.* FEMA 495. Washington, DC: FEMA.

FEMA. 2014a. *FEMA's disaster relief fund: Overview and selected issues.* Washington, DC: FEMA.

FEMA. 2014b. *ICC reducing flood losses through the international codes coordinating building codes and floodplain management regulations.* 4th ed. Washington, DC: FEMA.

FEMA. 2016. *National disaster recovery framework.* 2nd ed. Washington, DC: FEMA.

FEMA. 2017a. *Declaration data base.* Washington, DC: FEMA.

FEMA. 2017b. *Disaster relief fund: Monthly report as of July 31, 2017.* Washington, DC: FEMA.

FEMA. 2017c. *Flood insurance, comprehensive reform could improve solvency and enhance resilience.* Washington, DC: FEMA.

FEMA. 2017d. *Natural disasters data base.* Washington, DC: FEMA.

FEMA. 2018a. *Energy sector mitigation opportunities.* FEMA DR-4339-PR. Washington, DC: FEMA.

FEMA. 2018b. *2017 Hurricane season—FEMA after-action.* Washington, DC: FEMA.

FEMA. 2018c. *Identified mitigation opportunities.* FEMA DR-4339-PR. Washington, DC: FEMA.

FEMA. 2018d. *Resilience framework. 2018 Sustainability report and implementation plan.* Washington, DC: FEMA.

FEMA. 2019a. *Emergency support function #14—Cross-sector business and infrastructure DRAFT.* Washington, DC: FEMA.

FEMA. 2019b. *National Advisory Council. Draft Report to FEMA Administrator.* Washington, DC: FEMA.

FEMA. 2019c. *National mitigation investment strategy mitigation framework leadership group.* Washington, DC: FEMA.

FEMA. 2020. *Grants management modernization project plan.* Washington, DC: FEMA.

FEMA. n.d.-a. *Disaster (over-)insurance.* Washington, DC: FEMA.

FEMA. n.d.-b. *Disaster recovery reform act of 2018.* Washington, DC: FEMA.

FEMA. n.d.-c. *Disaster risk reduction minimum.* FEMA Policy 204-078-2. Washington, DC: FEMA.

FEMA. n.d.-d. *Executive order 12148.* Washington, DC: FEMA.

FEMA. n.d.-e. *FEMA 2028–2020 strategic plan.* Washington, DC: FEMA.

FEMA. n.d.-f. *Hazard mitigation assistance (HMA).* Washington, DC: FEMA.

FEMA. n.d.-g. *National flood insurance program (NFIP).* Washington, DC: FEMA.

FEMA. n.d.-h *National resilience strategy framing document.* Washington, DC: FEMA.

FEMA. n.d.-i. *Public assistance required minimum standards.* FEMA Recovery Policy FP-104-009-4. Washington, DC: FEMA.

Health and Social Services Sector. Dynamic Project Presentation. September 6, 2018.

Lindsay, B. R., W. L. Painter, and F. X. McCarthy. 2016. *An examination of federal disaster relief under the Budget Control Act.* Rep. No. R42352. Washington, DC: Congressional Research Service.

McCarthy, F. X. 2014. *FEMA's disaster declaration process: A primer.* Rep. No. R43784. Washington, DC: Congressional Research Service.

McCarthy, F. X., and N. Keegan. 2009. *FEMA's pre-disaster mitigation program: Overview and issues.* Washington, DC: Congressional Research Service.

REFERENCES

CRS. 2019. *Introduction to the National Flood Insurance Program (NFIP)*. Rep. No. R44593. Washington, DC: CRS.

CRS. 2020a. *The disaster relief fund: Overview and issues*. Rep. No. R45484. Washington, DC: CRS.

CRS. 2020b. *What happens if the National Flood Insurance Program (NFIP) lapses?* Rep. No. IN10835. Washington, DC: CRS.

DHS (Department of Homeland Security). 1988. *Stafford Act: Public Law 100-707 and modifications*. Washington, DC: DHS.

DHS. 2002. *Homeland Security Act 2002*. Washington, DC: DHS.

DHS. 2016. *Draft interagency concept for community resilience indicators and national-level measures*. Washington, DC: DHS.

DHS. 2019. *Shaken fury*. Washington, DC: DHS.

FEMA. 2011. *National preparedness system*. Washington, DC: FEMA.

FEMA. 2013. *Hurricane Sandy, FEMA after-action report*. Washington, DC: FEMA.

FEMA. 2019. *National response framework*. Washington, DC: FEMA.

NIBS (National Institute of Building Sciences). 2019. *National Institute of Building Sciences issues interim report on the value of mitigation*. Washington, DC: NIBS.

Painter, W. L., and J. T. Brown. 2017. *Congressional action on the FY2013 disaster supplemental*. Rep. No. R44937. Washington, DC: Congressional Research Service.

Smith, A. B. 2019. "2018's Billion dollar disaster in context." https://www.climate.gov/news-features/blogs/beyond-data/2018s-billion-dollar-disasters-context.

CHAPTER 5

ASSET AND SYSTEM MODELING CONSIDERATIONS FOR ASSESSMENT OF CIVIL INFRASTRUCTURE RESILIENCE

C. Mullen, R. Grant

5.1 DEFINITION OF RESILIENCE

The organization of the chapter is based on a presentation at a short course organized by the chair of the Objective Resilience Committee that was delivered on November 1, 2018, as a 2018 EMI International preconference offering on the campus of Tongji University in Shanghai, China (Ettouney, Lynch, and Mullen, unpublished report, 2018).

For the purposes of contextualizing this chapter, a strict definition of resilience is not required. However, the overarching theme adopted here relates to the definition provided in a previous chapter on the four R components of resilience. Implications of these four Rs were discussed in a breakout session of a workshop on Infrastructure Objective Resilience (Mullen 2018) that further refined these four Rs in the context of achieving Robustness/Hardening, Resourcefulness/Resource Management, Recovery/Duration, and Redundancy as a basis for setting objective metrics for resilience. How these are applied and interpreted will vary depending on the stakeholders including researchers, owners, and their representatives, engineers and architects, emergency managers, professional organizations, manufacturers and vendors, federal, state, and local governments, and the general public. One of the software tools discussed in the chapter will implicitly adopt a specific definition consistent with the four Rs to apply a specific numerical metric called a Risk Index.

5.2 RESILIENCE FACTORS FOR CONSTRUCTED FACILITIES

5.2.1 Constructed Facilities as Assets—Impact on System Resilience

The role of a constructed facility in a system within the context of resilience is essentially that of an asset whose function is key to the operation of the system. Thus, an assessment of the performance of the facility as an asset must focus on the loss of functionality in relation to the system or the potential reduction of the system performance as a result of damage to the asset caused by a disaster.

Consider the civil infrastructure assets shown in Figure 5-1.

Each of the facilities shown serves an important function to the local and regional communities served. Each facility, locality, and region is exposed to multiple hazards and has experienced a major disaster

Figure 5-1. Examples of constructed facility assets in civil infrastructure (photo credits: C Mullen): (a) Mario Cuomo Bridge over Hudson River, NY, (b) US 90 Biloxi Bay Bridge, MS, (c) Hudson Yards Development skyscrapers in Manhattan, NY, and (d) Maritime Museum, Biloxi, MS.

(Hurricanes Sandy and Katrina). Each plays an important role in the renewal and vitality of the communities and economies served.

Modeling of the complete system will be considered beyond the scope of this chapter. However, a recognition of the facility's role or impact on the performance of the system, be it transportation, political, financial, or economic, will be considered. The primary performance of the asset in this role will be assumed, as a constructed facility, to be physical including structural, geotechnical, and where relevant, geological. Identifying the role and relationship to the system is an important first step in identifying key characteristics that need to be incorporated in a model that assesses the performance of the asset and establishes an objective quantification of resilience.

5.2.2 Asset Functionality within a Civil Engineering System

Consider the two bridges shown in Figure 5-1. The large cable-stayed bridge (Mario Cuomo Bridge) serves as one of only a few major crossings of the Hudson River in New York. Further, its location on an interstate highway network carrying passenger and truck traffic into the northern portions of New York City (Bronx and Manhattan via I-87) as well as across counties just to the north (Westchester and Rockland via I-287) makes the bridge a key asset for both the metropolitan New York City area and the state. The more conventional girder bridge (US 90 Biloxi Bay Bridge) lies on a US highway that runs parallel to the major interstate artery (I-10) just to the north. The highway crosses the Biloxi Bay and provides local access from multiple Gulf Coast resort communities in the states of Mississippi and Alabama and one of the major entrances to the city of Biloxi where multiple casinos generate significant revenue for the state of Mississippi.

The two building cases provide contrasting examples of large and small redevelopment projects. The Hudson Yard Development is one of the largest real estate developments in the world enabling transformation of a part of Manhattan once considered an undesirable area of little relative importance. The Maritime Museum is a facility that was reconstructed after being destroyed in Hurricane Katrina. It has become a symbol of the resilience of the city and region in the aftermath of the devastating storm that is considered the costliest affecting the United States when inflation is taken into account. In the summer of 2019, an exhibit at the museum commemorated the 50th anniversary of another major storm, Camille—ranked 20th in the United States in terms of cost, and until Katrina the most devastating on record to impact the Mississippi Gulf Coast.

5.2.3 Critical and Valuable Facility Assets

A constructed facility may or may not be deemed critical to the performance of the system in which it serves as an asset. This is in contrast to the value of the asset itself. For the cases in Figure 5-1, each facility has significant value from a variety of contexts. Cost of course is a major context that affects the decision to replace the asset after a disaster should it be physically damaged. In the event the damage is more a loss of function, other factors come into consideration to bring the asset back into service. The value of the asset may not just relate to its expense in construction or reconstruction in the presence of damage. In the case of the bridge, there are direct and indirect values to be considered should the facility be out of service for any period of time because of it being part of a transportation system. In the building cases, the value can be related to occupancy or other real estate impacts, as in the case of the New York City Hudson Yards development project, or the recovery of community spirit, strength, and vitality, as in the case of the Biloxi Maritime Museum after a disaster.

A constructed facility may also serve as critical in the various stages of a postdisaster response and recovery. In the case of a bridge or highway, the actual operation of the facility may change because of its critical nature as a means of access to areas affected by the disaster. The critical aspect may be related either to the response within the affected region or as an enabler of operations just outside the affected area. In the case of the Hudson River bridge, the loss of function of bridges and tunnels owing to flooding from Hurricane Sandy made both the George Washington and Tappan Zee (name of the Mario Cuomo at the time) Bridges especially critical to allow both business and residential access and continuity of operations depending on vehicles. As the only crossings permitting truck traffic, this put unusual demands on these facilities.

Buildings become critical when they contain vital operations either in advance as a result of emergency management or business continuity planning or ad hoc as a result of special needs arising after the impacts of the disaster have become clear. The design of a facility deemed critical may or may not incorporate the protections needed to serve the required functions during or after a disaster depending on when the decision was made to make the facility a critical asset. When the decision is after the disaster strikes, the performance is severely impacted.

In Hurricane Katrina, for example, the state emergency management agency did not have the proper size or technological capabilities in its Emergency Operations Center (EOC) needed to respond to the unprecedented scale of the disaster. During the storm, casinos built of reinforced concrete that survived albeit with heavily damaged zones served as Forward Operating Bases (FOB) for the recovery. Many police and fire stations, hospitals, municipal buildings, and even military bases were not designed to resist the wind forces and/or surge generated by the storm and these facilities typically considered critical became hazards themselves (FEMA 2006).

5.2.4 Critical Asset Components and Characteristics

For each constructed facility type, there will be a number of components, component types, or characteristics that are predictably important to the performance of the facility. Critical asset features have been classified in a number of documents based on the facility type.

For building assets, this effort has been driven by agencies involved in seismic design (IBC 2018, IEBC 2018), postearthquake damage assessment (ATC 2005), and predisaster rapid visual assessment screening (ATC 2016). Some level of consistency has, thus, been established for this hazard type. These classifications have some carryover to the hurricane hazard (ATC 2004) with suitable additions that are hazard-specific for wind-induced damage.

For bridge assets, this effort has been incorporated in design (AASHTO 2018) and inspection (FHWA 2012) manuals. Classifications are also found in the national inventory (FHWA 2018) and modeling information standards (FHWA 2016).

A sample of critical asset components for buildings and bridges may be observed in Figure 5-2.

Efforts that attempt a consistent approach to multiple hazards and facility types are found in emergency management loss estimation and resilience index–based rapid visual survey tools (FEMA 2009, DHS-S&T 2011). Elements of these tools draw from design code terminology and load path concepts.

Figure 5-3 shows select examples of facilities identified as critical in scenario-driven multihazard loss estimation (Miah et al. 2014) and bridge vulnerability (Mullen 2013) studies. These examples highlight a number of facility types deemed critical from emergency management and economic sustainability viewpoints.

As implied previously, any set of criteria for deciding which asset components or characteristics are critical for establishing resilience in an objective manner must consider both the facility performance as an asset and its role in the performance of the system in which it operates. For bridges and buildings, the roles may change over the time frame of planning, response, and recovery. These roles may also change as a result of the perhaps unforeseen nature and extent of devastation associated with a particular disaster.

Assumptions must be made, of course, regarding the latter, for the purposes of planning. However, these assumptions must adapt to postevent ground truth. For example, planning studies and results of scenario loss estimation analyses performed in the EOC during emergency response preparations activated by Hurricane Dennis (Dennis) just a month prior to Hurricane Katrina (Katrina) showed the potential devastation of a direct strike of a Category 5 event to the coast of Mississippi. As it happened, the

(a)   (b)

(c)   (d)

Figure 5-2. Highway bridge assets—critical components and characteristics (photo credits: C Mullen): (a) main span—Mario Cuomo Bridge, NY, (b) approach span—Mario Cuomo Bridge, NY, (c) approach span—Biloxi Bay Bridge, MS, and (d) girder restraints—Biloxi Bay Bridge, MS.

storm made landfall in the neighboring state of Alabama, so no intense response preparations took place in Mississippi.

Similar scenario analyses were then performed live in the EOC during Katrina using periodic downloads of updates on the storm's actual path and strength. Both sets of runs (Dennis and Katrina) made with the loss estimation software, however, failed to capture the full scale of the devastation caused by Katrina because neither a wind plus surge hazard scenario was available at the time nor were the casinos that dominated the economy of the impacted coastal region included in the facilities' database. From the economic recovery point of view, these proved to be major oversights in estimation of resilience. Ad hoc adjustments were made at the time using digital elevation models imported live during the event to estimate the extent of flooding, which showed a significant number of schools in the area inundated as a result of the unprecedented surge heights.

Further, facilities deemed critical by the software as shown in Figure 5-3 did not prove up to the task. The storm surge isolated many of them, the

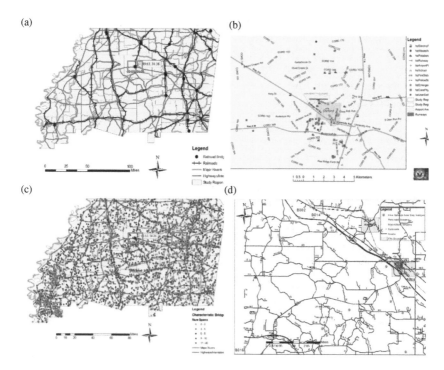

Figure 5-3. Critical facility assets in the north Mississippi region (plots generated using Level 1 database in 2009 HAZUS-MH): (a) bridges near the University of Mississippi, (b) disaster response facilities near the UM campus, (c) bridges near the UM campus and a major manufacturing plant, and (d) detail showing bridges closest to the plant.

wind caused damage to others, and in some cases, flooding rendered useless the equipment they depended on for functionality.

A similar argument could be made for the potential influence of an earthquake impacting the manufacturing facility shown in Figure 5-3. Loss of the facility for any extended period of time owing to physical damage of the buildings or loss of functionality of the highways supporting the facility's supply chain and product exports could have a major impact on the economy of the region and state and, thus, resilience.

5.3 PERFORMANCE MEASURES FOR CONSTRUCTED FACILITY RESILIENCE

5.3.1 Limit States, Damage Indices, and Loss Functions

Constructed facilities' performance by nature depends on physical characteristics in addition to operational features of what is housed in or attached to them. The physical characteristics include those mentioned in the preceding section. The performance measures applicable to them are well-documented in the resources identified previously and include material, element, subsystem, and facility-oriented structural and nonstructural system limit states.

For the typical construction materials used for bridges and buildings shown in Figures 5-1 and 5-2, important limit states include the following:

- Steel yield and ultimate stress;
- Concrete unconfined and confined ultimate stress;
- Section fully plastic moments;
- Maximum story drift ratios; and
- Plastic collapse and buckling loads.

Many of these limit states at the material and element levels are outlined explicitly in design codes and the consensus standards they reference. Figure 5-4 shows how they relate to potential damage scenarios. Some of

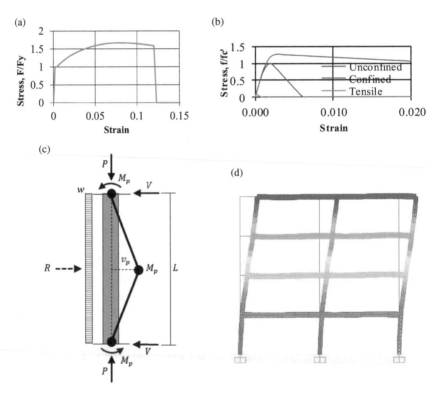

Figure 5-4. Constructed facility performance—material and element limit state characterization: (a) steel response, (b) concrete response, (c) beam-column plastic collapse, and (d) frame buckling mode.

these will be examined in detail later in the discussion of modeling and simulation.

At the subsystem and facility system levels, simpler damage states (slight, moderate, extreme) are defined so as to be more generally applicable. Definition of the states depends on the system type. A variety of such approaches are adopted in loss estimation, risk management, and rapid visual assessment software tools (*HAZUS-MH, IRVS*). These tools introduce additional performance measures such as damage indices, fragility curves, and loss functions to aid in the execution of the computational methodology. The former measures relate to the overall physical performance of the asset in global terms using key responses that may be monitored and quantified for a given event. The latter measure relates to the consequence of key responses of the asset in economic or functional terms.

5.3.2 Risk Factors and Resilience Indices

When considering modeling of assets and their system interactions and dependencies, it is important to consider the hazards (natural, such as earthquake, wind, or flood) or threats (man-made, such as terrorism) to which the asset may be exposed whether directly by the event or through the system interactions. Consequences of this exposure then need to be envisioned and the possible responses of the asset and its components that may be readily modeled objectively using the performance measures described previously.

Selection of hazards and consequences is a key part of the analysis of risk and consequently resilience owing to the broad scope of each. A broad set of risk factors may be automatically identified (*HAZUS-MH, IRVS*) in software-based analysis. When software is not relied on entirely for an analysis, possibly because of software limitations relative to analysis needs, a simpler set of factors may be preferred to provide a practical result (Swann and Mullen 2006).

Some software like *IRVS* define a resilience index as a numerical aid in decision-making. The index may make use of the selected risk factors and other data. Definition of resilience indices for special applications remains an active area of research.

5.3.3 Nonstructural Response, Human or Social Factors

As stated previously, the focus of this chapter is on physical response including structural and geotechnical primarily. Nonetheless, sometimes the physical response leads to a failure of nonstructural components supported or protected by the structural or geotechnical system. An important example of this is the fire suppression system. In the case of an earthquake or large blast, the well-designed structural system may sustain damage but still remain functional or repairable. Nonstructural components may, however, be more significantly damaged if not capable of sustaining large deformations or partial collapses of components or subsystems supporting them. Should a fire break out after the damaging event, the consequence in terms of functionality of the asset may prove catastrophic.

The linkage of the asset to the system may also prove to be a source of loss of functionality or resilience as a result of the scope of the disaster. Several major hurricanes (Katrina, Maria, Michael) have demonstrated that well-designed buildings including homes and even hospitals that successfully resisted the impact of the damaging event became essentially nonfunctional because devastation around the buildings either prevented access to them or degraded their personal property or social value.

5.3.4 Response, Repair, Restoration, Recovery

Quantifying resilience of an asset objectively through analysis or simulation will necessitate that the tools used to do so quantify the assets' performance through all stages of the event through recovery. Where the use of resilience indices is made, the levels of performance require correlation to the value of the index.

Loss estimation tools (*HAZUS-MH*) often include aspects of this correlation through incorporation of limit state–based fragility curves coupled with damage-related loss functions. The direct relationship to a resilience index remains unquantified in this case. Further such tools are typically hazard-specific and scenario-driven. The multihazard or hazard-independent resilience is not addressed in this case either.

Rapid visual assessment tools (*IRVS*) that do incorporate a resilience index and multiple hazards do not rely on detailed simulation of the response to specific hazards but rather, matrix-driven correlations that depend on broad ranges of data that predict behavior attributed to the asset under consideration from a knowledge of construction details and visual observations. While not as rigorous as a detailed asset simulation, the ability to initiate analysis quickly and at a relatively low cost is a valuable quality of rapid assessment tools. Covering a large number of assets with a preliminary assessment can lead to better decisions regarding which assets need more detailed analysis and allow for a higher average level of evaluation for wider evaluations.

Ideally, a detailed asset simulation as might involve the use of advanced finite-element analysis tools (e.g., SAP 2019, ABAQUS 2019) might be performed for the purposes of resilience index quantification. The modeling should include not only the representation of the hazard or threat in terms of direct physical impact on the asset, but also the idealization of the physical response considering any deteriorated conditions of the asset at the time the loading is applied. For some critical assets, the modeling might even necessitate a consideration of fluid–structure or soil–foundation–structure interaction. The implications for repair and restoration of the asset to full functionality should then be assessed using repair/reconstruction cost information appropriate for the time period and region where the asset is located. The linkage of the asset to the system would also require a consideration for a final determination of the resilience index.

5.4 SIMULATION TOOLS FOR ASSET AND SYSTEM RESILIENCE ASSESSMENT

5.4.1 Finite Elements

Capturing the detailed physical response of an asset usually requires an accurate representation of the geometry of the structural, geotechnical, and sometimes geologic features of the asset as well as the important material properties and engineering analysis solution algorithms needed to define and relate the loading, boundary conditions, and response quantities of interest. When the quantities are mechanical such as motion and stress/force or thermal such as temperature or heat flux, a general-purpose finite-element (FE) software (e.g., *SAP2000* or *Abaqus CAE*) will provide the needed capabilities for simulating constructed facilities subject to the effects of natural hazards or man-made threats.

Figure 5-5 shows three-dimensional physical FE models of bridges and buildings demonstrating some of the aforementioned basic features. All but the 4-story building model were created with a general-purpose FE tool (*Abaqus CAE*) that employs advanced topology features for the creation of the analysis model and is oriented toward solid modeling. The 4-story building model was created with an FE tool (*SAP2000*) that emphasizes the use of templates for the creation of the analysis model, and while oriented toward structural design, the tool is useful for quantifying the response of typical constructed facilities subject to a wide range of hazards if the effect of the loading can be established on the members.

The FE models shown in Figure 5-5 are capable of capturing complex response behaviors. Nevertheless, considerable skill is required on the part of the analyst to interpret and capture an accurate representation of the loading event and the responses key to an objective determination of resilience.

A more complete description of the details of the modeling and simulation for several building cases relating to the models shown in Figures 5-4 and 5-5 may be found in Sideri et al. (2017) and Dahal et al. (2018).

(a)

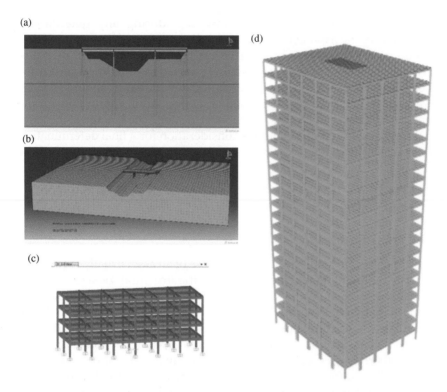

(d)

(b)

(c)

Figure 5-5. Constructed facility physical modeling—FE models: (a) geometry/ topology for a geology–soil–foundation–structure interaction (GSFSI) model of a 3-span bridge for a scour hazard study, (b) bridge scour GSFSI model showing FE mesh, (c) 4-story building frame for a hazard-independent stability study , and (d) 20-story building for a blast collapse vulnerability study.
Source: (a) Mullen and Powell, unpublished report, 2016, (b) Mullen and Powell, unpublished report, 2016, (c) Mullen and Dahal, unpublished report, 2018, Mullen and Dahal, unpublished report, 2017.

5.4.2 Geospatial Analysis

The linkages of the asset to a system and its surrounding regions are often best captured using a representation of the geospatial locations of other assets and important features of the loading event. Such a representation enables the propagation and attenuation characteristics of the dynamic event to be captured for assets in their respective positions relative to the source.

Figure 5-6 shows a range of disaster scenarios whose intensities at critical facility locations may be estimated using physics-based modeling or heuristic reasoning. The scenarios shown include earthquake ground

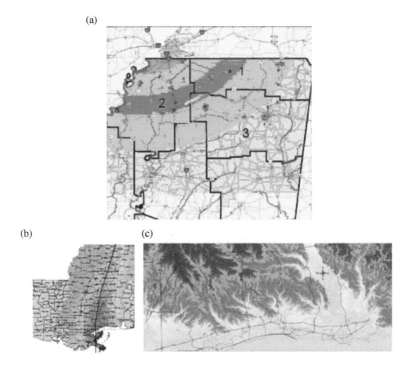

Figure 5-6. Geospatial hazard modeling of constructed facilities in disaster scenarios: (a) earthquake peak ground acceleration in north MS (the heavy black lines show Mississippi DOT Districts), (b) hurricane wind speeds in MS (the heavy black line shows an assumed path of the eye), and (c) hurricane surge impact in coastal MS (using digital elevation model).
Source: Plots generated using 2005 HAZUS-MH.

motion and hurricane wind speeds and surge flood elevation levels. Other studies in which this approach has been implemented include tornado (DHS-S&T 2011) and blast (Tadepalli and Mullen 2013) event scenarios. The visualization and quantity estimation enable a more realistic interpretation of these extreme events and their consequences. More sophisticated modeling can enable simulation in real time.

5.4.3 Artificial Intelligence

Because accurate and complete 3D models and databases of the performance of constructed facilities during disasters become available and probabilistic methods of analysis of response integrated with 3D models become more powerful and widely used, the opportunities for improving response estimation for a variety of situations through the use of artificial intelligence increase. Detailing these opportunities and their role in improving the resilience of civil infrastructure is beyond the scope of this effort.

5.5 ASSET RESILIENCE—THE ROLE OF SIMULATION

5.5.1 Multihazard Asset Simulation

The goals and nature of the objective resilience simulation of a constructed facility asset vary depending on whether the outcome is to be the design of a new facility, the estimated performance of an existing asset or a group of assets in the event of a hypothetical or active disaster scenario for emergency management planning, or a detailed evaluation of vulnerability of an existing asset for repair or strengthening decisions in a deteriorated or postevent damage condition.

In design, the primary goal is to achieve performance that exceeds consensus criteria given in terms of prescribed limit states (AASHTO, 2018) using analysis methods that also meet consensus expectations. A key part of the process is to use methods of determining minimum loads established by consensus in combination with other loads combined in a manner also established by consensus.

Although the load combination implies a multihazard approach, the sequence of events is not considered explicitly. Also, the use of simulation to estimate sometimes complex responses often needed to quantify the performance with respect to limit states may or may not be required. Performance-based design (PBD) is currently explicitly used only in consensus standards and code provisions for existing facilities except in special circumstances. For example, some large metropolitan areas with a concentration of tall buildings have voluntarily adopted PBD for new buildings.

Recently, a software tool (ASCE 2019) that implements the consensus standard approach adopted by many building codes in the United States has been made available to the design community. Figure 5-7 shows an example of output from the tool.

5.5.2 Multihazard System Performance Simulation

As has been mentioned, it is useful to establish the performance of the linkage of the asset to the system, as well as that of assets that are in the vicinity of the asset of interest. Hazard simulations that are defined by scenarios that are also modeled geospatially enable the latter. Modeling the linkage to the system is naturally handled this way as long as the linkage is a physical one that requires connectivity through other assets in the vicinity of the asset of interest and the hazard. For example, bridges on the

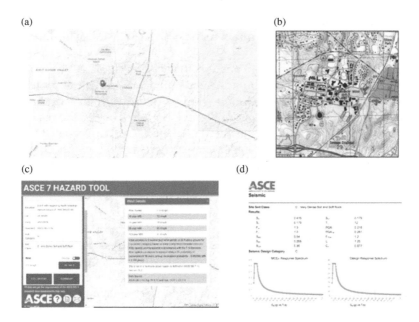

Figure 5-7. Tool for estimating multihazard extreme loads for design (using 2019 ASCE Hazard Tool): (a) menu of loads and location view for the UM main campus, (b) GIS map locating campus facilities, (c) seismic hazard design parameters for the main campus site, and (d) seismic hazard design parameters for the site.

same highway system would be physically linked to one another and performance will depend on both that of the other bridges in the network and that of the roadways connecting them.

System simulation in the bridge network case can initially be considered through the separate physical response of each asset to a hazard such as an earthquake that impacts a large region. The ground motion simulation provides the important intensity measures at each bridge site separately based on distance from the source and attenuation models. The response of each asset is then considered based on the characteristics of the asset given the intensity at the site. This can be done in a simplified way through intensity–response relationships for basic categories of the asset or can be done through more advanced FE techniques if sufficient information is available for the asset.

5.5.3 Case Studies

Constructed facility performance observations and subsequent evaluations are available from a number of hurricanes, floods, and earthquakes that have struck communities in the United States and around the world over the last two decades. Here, we draw on two case studies performed by the lead author where modeling and simulation were performed and implications drawn for objective resilience determination.

5.5.3.1 Case Study 1—Resilience of Bridges in North Mississippi.

Hurricane Katrina severely impacted the built environment in a number of ways through its devastating combination of high winds and unprecedented surge. A number of highway bridges key to the resilience of coastal communities in Alabama, Mississippi, and Louisiana (AL, MS, and LA) were destroyed. This experience demonstrated the importance of these lifelines and the hazard posed by large-scale events on an aging infrastructure.

Although emergency planners in MS had been concerned by the potential impact of a catastrophic earthquake for some time (Swann and Mullen 2004), the firsthand experience of Katrina raised fresh concerns over the ability of the aging infrastructure in north MS to perform adequately. A major federally sponsored study (Elnashai et al. 2009) had clearly shown the magnitude of the impacts of an M7.7 event on the New Madrid Fault over an 8-state region that included MS using the GIS-based methodology described previously.

To prepare for a multistate exercise for the potential occurrence of the M7.7 eventually included in formal response plans of the eight affected states, MS planners decided that an FE-based simulation approach would provide a clearer picture of what to expect for bridges in the state. Three bridges crossing an important river cutting through the affected part of the state and would prove critical to both response and recovery plans were identified for further study.

A combination of 2D and 3D soil–foundation–structure interaction models was used to establish the performance of these bridges with respect to the critical limit state of formation of a plastic collapse mechanism of the pile-supported bent structures in both pushover and simulated time history analyses (Mullen 2011). These analyses focused on the use of structural-oriented models and lumped property equivalent elements to capture soil deformation response.

Subsequent to the aforementioned studies, a federally sponsored study enabled the importance of deterioration on simulated response to be investigated using more advanced FE models of select bridges in north MS adopting a solid modeling approach (Mullen and Powell 2016). The first set of models and analyses focused on the ability of modal analysis–based measurements to capture the effect of material deterioration on stiffness degradation (Mullen and Dahal 2017).

Subsequent modeling and analyses in the same study were able to use the more detailed approach to examine the more-difficult-to-capture effects of severe scour on the stiffness degradation and global stability for one of the bridges located in north MS that is key to economic development owing to its proximity to a major manufacturing plant.

Figure 5-8 shows selected results of the more detailed FE modeling approach and the highly three-dimensional (3D) behavior of the bridge in the presence of severe scour.

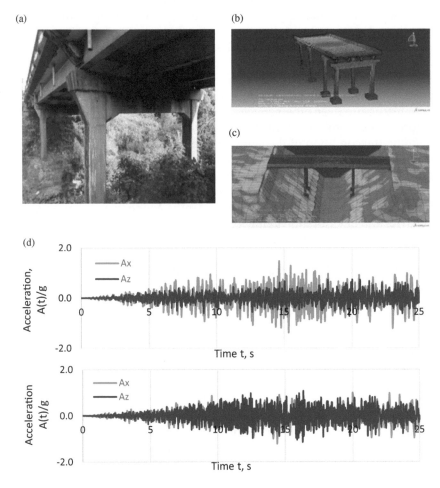

Figure 5-8. Solid modeling FE simulation for a response evaluation of an aging bridge (plots generated using Abaqus CAE): (a) view showing existing conditions at the time of scour study. (b) mode shape for a fixed-based model, (c) mode shape of a full model with soil and symmetric scour between intermediate bents, and (d) seismic response simulation results for a full model with soil, with and without asymmetric scour.

Source: Mullen and Dahal (2017). Photo credit: Chris Mullen.

A number of factors turned out to influence the response of what appears to be a rather normal bridge in a rural setting. First, although not shown here, the exactly same design of the superstructure and multicolumn bents, a typical situation at the time of construction in the mid-1930s, was applied in two very different sets of site conditions. While both crossed streams, one bridge lay on a severe skew over the crossing, whereas the other had none (Figure 5-8). One lay on soft soil requiring piles, whereas the other lay on a chalk layer requiring only spread footings. Further, the natural and man-made interventions on the flow conditions led in one case to a deepening of the bed between the piers with severe scour around one of the piers and buildup of material around the other.

Another factor influencing the response of the bridge is the apparent concentration of material degradation in certain parts of the superstructure or substructure over others. Such deterioration leads to uneven stiffness degradation. Combined with the scour variability, the issue of properly defining the condition of the aging bridge for the purposes of devising a suitable model for resilience determination becomes a challenge without a great deal of site-dependent characterization.

Figure 5-8 shows that, for the bridge model studied, these factors can influence the dynamic response significantly. Mode shapes involve 3D behavior and deterioration influences certain modes over others depending on the distribution of the deterioration. Vulnerability to extreme loading is increased by the amplified motions in these preferential modes caused by the deterioration. In the example shown in Figure 5-8, the scour pattern completely changes the relative importance of one component of motion over another.

5.5.3.2 Case Study 2—Resilience of Buildings on the UM Main Campus. The influence of a major earthquake on buildings in north MS has also been a major concern. A number of select facilities were identified for detailed FE simulation prior to Katrina. These included a hospital and several UM campus buildings (Hackett et al. 1997, Mullen and Swann 2001). Just prior to Katrina, a more comprehensive look at critical facilities in north MS was undertaken using the geospatial and database-driven approach (Mullen and Desai 2005). As it turns out, a facility-specific database for the UM campus did not exist in the software, so a campus-specific study was undertaken (Swann and Mullen 2006).

Katrina (FEMA 2006) demonstrated that the impact of a disaster can be significantly affected by the variable adoption of building codes by different communities, from county to county, from city (urban) to surrounding county (rural), and from private to government owned. Also, a state law regulating gambling led to unforeseen impacts not only on casino buildings built on moored barges but also on the hotels serving them that were located close to them and, thus, highly exposed to the forces of both wind and surge. Much is to be learned then about the linkage of the resilience of a facility to others in a system or community. Also, buildings are affected not only by other buildings but also by the transportation and other infrastructure systems including cyber serving them.

The UM campus study provides a microcosm of the variable influences on the performance of buildings in a community. Built in 1848, the building stock varies widely in construction technology and age. Historic unreinforced masonry buildings (Mullen et al. 1997) are still actively used. Numerous aging reinforced concrete buildings housing classrooms, offices, laboratories, and dormitories, including some with post-tensioned floors, were built in the last century. This construction comprises design based on a variety of building codes not recognizing the earthquake hazard. Recent construction built to more modern standards that does consider the earthquake hazard also includes steel frame structures housing sports arenas and laboratories. New residence halls have been built with cold-formed steel wall and roof systems. Some older brick facades that have deteriorated badly have been replaced with architectural precast elements, and new parking garages are built with prestressed precast concrete elements.

Figures 5-6 and 5-7 show some plan views of the geospatial view of many of the facilities. Modeling of the response of the buildings to natural hazards was studied first in a federally sponsored study that used an early version of software created under the direction of the *NIBS* premised on a probabilistic loss estimation methodology (FEMA 2000). The software was only capable of modeling the earthquake hazard at the time. Subsequently, the hurricane and flood hazards were added (FEMA 2009), the latter requiring the incorporation of digital elevation models (DEM) to establish the flood depths at facility locations. The latter feature was not yet available for wide use at the time of Katrina, and, therefore, a manual overlay approach was used to create the map shown in Figure 5-6.

The first study provided the key quantitative basis for the development of the first federally approved natural hazard mitigation plan for the campus (Swann and Mullen 2006). Because no geospatially referenced database for the facilities on the state's inventory existed, the first plan required this be developed manually along with an interpretation of reconstruction costs that had to consider the varying times of construction and cost records in the inventory.

A second study was later undertaken to update the plan (Miah et al. 2014). In the interim since the first study, new information led to a reconsideration of both the hazard exposure and the asset inventory. A moderate tornado struck near the campus and major tornados struck a small town near the campus as well as nearby states, including the campus of the University of Alabama. Also, the University of Mississippi underwent the largest growth spurt in its history with a corresponding expansion of its housing, student services, and athletic facilities.

During that time, the understanding of objective resiliency was developing and the IRVS risk factor and resilience index evaluation tool became available. While the tool appeared promising as a way to identify relative asset risk and resilience to support investment decisions, a lack of experience with the tool and its formulation prevented its use in the study as the basis of the updated plan that was subject to state and federal approvals for acceptance and implementation over the next five-year update cycle. A limited effort was made, however, to explore the tool's application to a dozen critical and high-priority facility assets to evaluate its potential use in future evaluations that were not as time- and resource-constrained. Representative evaluations are shown in Figure 5-9.

The two facilities represent not only two important assets on the campus; they also represent major operational functionalities whose resilience during a catastrophic event will be key. The towers comprise the largest concentration of full-time on-campus housing, whereas the power plant maintains the independence of the campus physical plant from the regional power grid. The towers will, in fact, be heavily dependent on the power plant's continuous operation after a damaging event. A formal study of the degree of this dependence and postevent operational system and the impact on the resilience indices shown in Figure 5-9 has not yet been evaluated. The evaluation of network resilience has been identified as a needed extension of the current IRVS single asset focus.

Noteworthy outcomes of the results in Figure 5-9 as isolated assets include the dominance of tornado risk over earthquake and a relatively high overall resilience index. Note that the twin dormitory towers exhibited an especially high tornado risk, whereas the power plant facility exhibits a noticeably lower earthquake shaking risk and a slightly higher overall resilience. It should be noted that the measurements of resilience in the IRVS tool are relative and become more valuable when compared with other buildings of similar types as opposed to being used as absolute values. As isolated measures, they can, as noted here, be used as indicators or potential problems needing to be addressed to make the facility relatively safe and secure.

At 10 stories, the towers are the tallest buildings on the campus. They were built around 50 years ago when no building code had been formally adopted by the state for university buildings. Even if they had been adopted, the codes existing at that time would not have required any special design provisions for earthquake resistance. By contrast, the power plant is housed in a low-rise building constructed in the last 10 years by which time codes began including such provisions.

Only in 2005 did the state adopt modern building codes in a formal policy statement governing all new university facilities. By then,

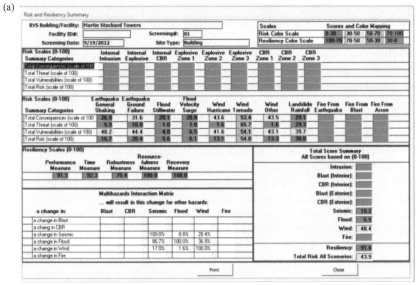

50-year old multistory dormitory

new critical equipment facility

Figure 5-9. Risk factor and resilience index simulation for campus assets (plots generated using IRVS): (a) 50 year-old multistory dormitory, and (b) new critical equipment facility.

earthquake design provisions had been a major consideration whose demands at the campus location would in most cases exceed those of thunderstorm-related winds.

While the year of building code adoption might be misconstrued as a result of the experience from Hurricane Katrina, it was, in fact, accomplished before the event and as a more direct consequence of the recognition by the state's higher education governing policy body of studies completed earlier that year. These included the first UM campus plan (Swann and Mullen 2006) and the first state Standard Mitigation Plan (Swann and Mullen 2004) incorporating an earthquake hazard component.

The aforementioned comments focus on the physical infrastructure aspects of resilience that are recognized in the HAZUS-MH and IRVS methodologies. Social, economic, and emotional impacts not addressed by these tools are now being recognized as important to assess. A new interdisciplinary UM faculty research focus group (University of Mississippi 2019) was formed in 2018 and recognized by the university administration to study such impacts across these various dimensions.

5.6 BEST PRACTICES

This chapter promotes the role of modeling and simulation of assets and the systems in which they function in enhancing the objective assessment of resilience of such assets and systems. Examples and case studies of building and bridge assets and systems are discussed to highlight the key characteristics influencing such an assessment and aspects that are clearly demonstrated, noting that some of these have yet to be explored fully and will depend on continuous improvement in the state of knowledge, the development of modeling and simulation tools incorporating this improved understanding, and the training and implementation of the professional community in best practices.

The continued development of accurate and complete Building Information Models (BIM) of facilities (buildings and infrastructure) that capture system visual as well as operational characteristics will make these analyses more realistic and their results more valuable and able to be validated. Continued improvements in the integration of relevant aspects of BIM with geospatial information will allow for a more thorough and realistic evaluation of interdependencies between assets.

From the discussion provided in this chapter, the following best practices for objective resilience–related modeling and simulation of infrastructure assets such as buildings and bridges are suggested to be utilized and further developed to the extent to which project time and computational resources allow.

Asset-Level Resilience

1. Identify the key characteristics of the asset components and subsystems in isolation; this is vital to sustaining normal physical and operational performance.
2. Examine loss and recovery measures for such performance applicable to the key characteristics using either a multihazard or a hazard-independent approach.
3. Establish scenarios with multihazard sequences as appropriate that incorporate physics-based hazard intensity and asset and geospatial modeling and simulation.
4. Apply multidimensional modeling and simulation as needed to evaluate asset performance measures incorporating as many of the elements of objective resilience as may be feasible.

System-Level Resilience

1. Establish updated inventories of critical assets comprising a system or interdependent group of assets defined by geospatial proximity, hazard exposure during scenario events, operational functionality, and/or organizational authority.
2. Establish relationships between the interdependent assets as defined by relative and sequential impacts on disaster resilience of the group.
3. Establish the effects of critical facility component and subsystem performance loss on interdependent asset performance.
4. Evaluate the effects of scenario events on system and asset group performance including critical facility components and subsystems and interdependencies.

REFERENCES

AASHTO (American Association of State Highway and Transportation Officials). 2018. *AASHTO LRFD bridge design specification.* 14th ed. Washington, DC: AASHTO.

ABAQUS. 2019. *Abaqus CAE: Users manual,* version 6.19. Providence, RI: Simulia.

ASCE. 2019. *ASCE hazard tool.* New York: ASCE.

ATC (Applied Technology Council). 2004. *Safety evaluation of buildings after windstorms and floods.* ATC-45 Field Manual. Redwood City, CA: ATC.

ATC. 2005. *Field manual: Postearthquake evaluation of buildings.* 2nd ed. ATC 20-1. Redwood City, CA: ATC.

ATC. 2016. *Rapid visual screening of buildings for potential seismic hazards: A handbook.* 3rd ed. ATC/FEMA P-154. Redwood City, CA: ATC.

Dahal, P., T. Powell, and C. Mullen. 2018. "Hazard-independent stability sensitivity study of steel and RC frame structures." *J. Civ. Eng. Constr.* 8 (2): 63–69.

DHS-S&T (Department of Homeland Security Science and Technology Directorate). 2011. *Integrated rapid visual screening of buildings.* Buildings and Infrastructure Protection Series, BIPS 04. Washington, DC: DHS-S&T.

Elnashai, A., T. Jefferson, F. Fiedrich, L. Cleveland, and T. Gross. 2009. Vol. 1 *of Impact of New Madrid seismic zone earthquakes on the Central USA.* MAE Center Rep. No. 09-03. Urbana, IL: Mid-America Earthquake Center.

FEMA. 2000. "HAZUS 99." Accessed November 1, 2021. http://www.disastersrus.org/emtools/earthquakes/FEMA366.pdf.

FEMA. 2006. *Hurricane Katrina in the gulf coast: Mitigation assessment team report-Building performance observations, recommendations, and technical guidance.* FEMA 549. Washington, DC: FEMA.

FEMA. 2009. *HAZUS-MH: Multi-hazard loss estimation methodology-earthquake module.* Technical Manual, Version MR5. Washington, DC: FEMA.

FHWA (Federal Highway Administration). 2012. *Bridge inspector's reference manual.* FHWA-NHI 12-049. Washington, DC: US Department of Transportation, FHWA and National Highway Institute.

FHWA. 2016. *Bridge information modeling standardization.* FHWA/HIF 16-011. Washington, DC: US Department of Transportation, FHWA and National Highway Institute.

FHWA. 2018. *National bridge inventory.* Washington, DC: FHWA.

Hackett, R., C. Swann, and C. Mullen. 1997. *Structural seismic vulnerability evaluation of Baptist Memorial Hospital-Desoto.* Final Rep. Central United States Earthquake Consortium, Department of Civil Engineering and Mississippi Mineral Resources Institute.

IBC (International Building Code). 2018. *IBC, International Code Council, Third Printing, 2019.* Washington, DC: IBC.

IEBC (International Existing Building Code). 2018. *IBC, International Code Council, Third Printing, 2019.* Washington, DC: IBC.

Miah, M., K. Bethay, and C. Mullen. 2014. "Multi-hazard risk analysis and resilience indices for critical facilities on the main campus of the University of Mississippi." *Homeland Secur. Rev.* 8 (1): 79–100.

Mullen, C. 2011. *Seismic vulnerability assessment of critical bridges in North MS.* Final Rep. Pearl, MS: University of Mississippi, Department of Civil Engineering.

Mullen, C. 2013. "FE based vulnerability assessment of highway bridges exposed to moderate seismic hazard." Chap. 8 in *Engineering seismology, geotechnical and structural earthquake engineering*, edited by S. D'Amico. London: In-Tech.

Mullen, C. 2018. "Infrastructure asset resilience-role of simulation." In *Infrastructure objective resilience workshop*. Ann Arbor, MI: University of Mississippi.

Mullen, C., and K. Desai. 2005. "Earthquake damage assessment and liquefacion potential in Northern Mississippi." In *Mississippi Academy of Sciences 69th Annu. Meet.*

Mullen, C., and T. Swann. 2001. "Seismic response interaction between subsurface geology and selected facilities at the University of Mississippi." *Eng. Geol.* 62: 223–250.

Mullen, C. L., S. R. Swatzell, and R. M. Hackett. 1997. "3D modeling of an historic masonry structure for rehabilitation planning and seismic retrofit." In *Structural studies, repairs and maintenance of historical buildings*, 299–308. Southampton, UK: Computational Mechanics Publications.

SAP. 2019. *SAP2000: Users manual*, version 21. Berkeley, CA: Computers and Structures.

Sideri, J., C. Mullen, S. Gerasimidis, and G. Deodatis. 2017. "Distributed column damage effect on progressive collapse vulnerability in steel buildings exposed to an external blast event." *ASCE J. Perform. Constr. Facil.* 31 (5): 04017077.

Swann, C., and C. Mullen. 2004. *The State of Mississippi standard mitigation plan-earthquake risk assessment*. Final Rep. Pearl, MS: Mississippi Emergency Management Agency, Center for Community Earthquake Preparedness.

Swann, C., and C. Mullen. 2006. *Natural hazard mitigation plan of the University of Mississippi, Lafayette County, MS*. Final Rep. Pearl, MS: Mississippi Emergency Management Agency, Center for Community Earthquake Preparedness.

Tadepalli, T., and C. Mullen. 2013. "Simplified blast simulation procedure for hazard mitigation planning." Chap. 13 in *Design against blast: Load definition and structural response*, edited by S. Syngellakis, 133–142. State-of-the-art in Science and Engineering volume 60, Southampton, UK: Transactions of the Wessex Institute.

University of Mississippi. 2019. "UM disaster resilience constellation." https://flagshipconstellations.olemiss.edu.

CHAPTER 6
RESILIENCE MANAGEMENT

Mohammed M. Ettouney

6.1 INTRODUCTION

6.1.1 Definitions

Natural and man-made hazardous events can impose a devastating cost to society. As Figure 6-1 shows, the costs of some of these disasters in the United States alone can be staggering. Stakeholders of civil infrastructure have a vested interest in reducing these costs by improving and maintaining operational and physical high performance through providing resilient systems and communities. We first need to define what resilience is and then discuss some of its main features.

6.1.1.1 The 4Rs. Infrastructure resilience has been defined in numerous ways, and a popular definition was introduced by NIAC (2009), which states

> Infrastructure resilience is the ability to reduce the magnitude and/ or duration of disruptive events. The effectiveness of a resilient infrastructure or enterprise depends upon its ability to anticipate, absorb, adapt to, and/or rapidly recover from a potentially disruptive event.

NIAC also determined that resilience can be characterized by three key features:

1. *Robustness*: The ability to maintain critical operations and functions in the face of crisis. This includes the building itself, the design of the infrastructure (office buildings, power generation, distribution

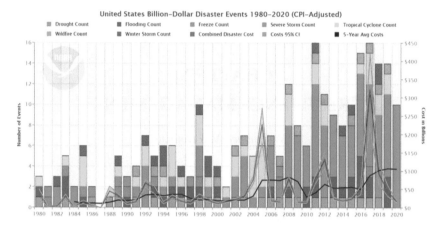

Figure 6-1. Frequency and cost of natural disasters in the United States.
Source: NCEI (2020).

    structures, bridges, dams, levees), or in system redundancy and substitution (transportation, power grid, communications networks).

2. *Resourcefulness*: The ability to skillfully prepare for, respond to, and manage a crisis or disruption as it unfolds. This includes identifying courses of action and business continuity planning, training, supply chain management, prioritizing actions to control and mitigate damage, and effectively communicating decisions.

3. *Rapid recovery*: The ability to return to and/or reconstitute normal operations as quickly and efficiently as possible after a disruption. Components [of rapid recovery] include carefully drafted contingency plans, competent emergency operations, and the means to get the right people and resources to the right places.

We propose that resilience has another key feature: *redundancy*. Sometimes, these four resilience features are simply called the 4Rs. From the 4Rs, it is clear that resilience is multidisciplinary: cooperation of different disciplines and agencies are needed to achieve resilient assets and communities. Without multidisciplinary cooperation and contributions, there cannot be sound or efficient resilient infrastructure.

    A beneficial conceptual illustration of resilience as a graph with recovery time-operational quality as its horizontal and vertical axes, respectively, was introduced by Bruneau et al. (2003). The graph is detailed in Figure 6-2. Figure 6-2(a) shows graphically how we can objectively estimate the resilience of an asset or community by utilizing resilience charts. We then use Figure 6-2(b) to show a comparison of the resilience of two assets (or communities). The illustration shows as an undesirable event impact two assets (or communities): The operations of asset (or community) "A" will

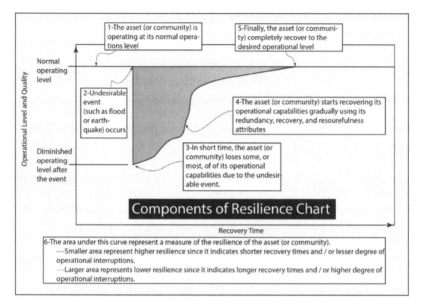

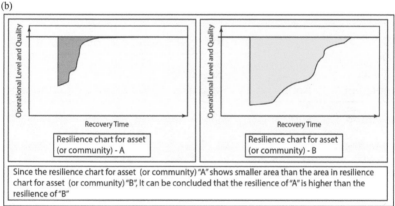

Figure 6-2. (a) Resilience definition in the time-operational quality space (showing the resilience chart), and (b) Comparison between the resilience of two assets (or communities) using resilience charts.
Source: Adaped fron Ettouney (2014).

immediately lose some operational quality, then start recovering until returning to full operational quality. As for asset (or community) "B," it will lose much of its operational quality when subjected to the same undesirable event. It will recover but on a much slower pace than asset "A." Thus, we can conclude that "A" is more resilient than "B." We can use the area under the curve that describes the time-operational quality behavior of the asset of the community to objectively assess the resilience of the asset of the community.

6.1.1.2 PPD-8 and PPD-21. Another popular resilience definition, PPD-8, was introduced by NSC (2011). It was then updated as PPD-21 by the Office of the Press Secretary (2013). As in almost all popular resilience definitions, its basic contents are similar to the 4Rs definition (see Gerasimidis and Ettouney 2022). However, it contains eight components instead of four. The components of each definition are fairly related, as shown in Table 6-1.

6.1.1.3 Resilience, Risk, and Sustainability. Before we end our discussion of resilience definitions, it is of interest to clarify the differences and relationships between resilience and another two important paradigms in civil infrastructure: risk and sustainability. Risk is often expressed as the relationship between a particular hazard (or threat) that might degrade the performance of the infrastructure under consideration and the consequences that might

Table 6-1. Relationship between Resilience Components of the Two Definitions.

| PPD-8/ PPD-21 resilience component | Relationship to 4Rs resilience components | | | |
	Robustness	Resource-fulness	Recovery	Redundancy
Prevention				
Preparedness	Yes	Yes	Yes	Yes
Protection/ Robustness	Yes			Yes
Asset versus Community treatment	Yes	Yes	Yes	Yes
Mitigation	Yes	Yes	Yes	Yes
Resource allocations/ planning	Yes (budgeting)	Yes	Yes (budgeting)	Yes (budgeting)
Response		Yes	Yes	
Recovery	Yes (reconstruction)	Yes	Yes	Yes (reconstruction)

result from a degradation of performance (see Gutteling and Wiegman 1996; FEMA 2005; NRC 2010). Most professional industries, such as engineering, finance, insurance, and medicine, adopt a variant of this particular definition of risk (Gutteling and Wiegman 1996). In the infrastructure community, FEMA (2005) uses an objective risk definition that states

$$Ri = Ri(T,V,C) \tag{6-1}$$

with Ri, T, V, and C representing risk, threat (hazard), vulnerability, and consequences, respectively.

Of course, the type of risk function depends on the desired degree of complexity of risk analysis.

We have seen previously that a reasonable resilience definition relates resilience to robustness, resourcefulness, recovery, and redundancy (the 4Rs). On further reflection, we can see that the 4Rs can be recast as a subset of C, T, and V. Ettouney and Alampalli (2017b) proposed a relationship between C, T, V and the 4Rs, as shown in Table 6-2. Another way to illustrate the subjective relationship between the components of risk, Ri, and resilience, Re, is described in Figure 6-3.

Based on the preceding discussions, we conclude that resilience can be recast as a special form of risk. We note that resilience focuses mainly on the continuity of operations or the degree and duration of operational interruptions. Thus, if we cast risk consequences, and the threat and vulnerability parameters that affect those two types of consequences, we will end up with exactly the very definition of resilience.

Sustainability is a bit more difficult to define because of the immense number of definitions. Some sustainability objectives include the following: operating within acceptable and renewable limits; allowing environmental, societal, economical, and humanitarian concerns govern actions; or having respect for available natural and universal resources (see Tehrani and Nelson (2022) for an in-depth discussion about the interrelationships between sustainability and resilience).

Table 6-2. Relationship between Risk and Resilience.

Risk components	Resilience components			
	Robustness	Resourcefulness	Recovery	Redundancy
Consequences	Minor	Major	Major	Major
Threat	Major	Minor	Minor	Major
Vulnerability	Major	Minor	Minor	Major

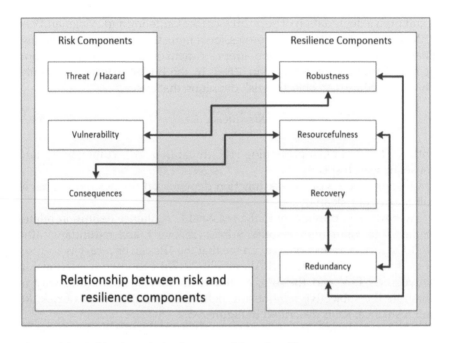

Figure 6-3. Subjective relation between risk and resilience.
Source: Ettouney (2014).

Reflecting on sustainability objectives, we can recast them in terms of a risk's threats, vulnerability, and consequences. Sustainability, with all of its varying interpretations, is also a subset of risk. Unfortunately, because there is no universal definition of sustainability components similar to resilience components, we cannot offer a table showing relationships between sustainability components and risk components as we have done for the risk–resilience interrelationships. Suffice to say, threats, vulnerabilities, and consequences in sustainability are a subset of risk application. Table 6-3 shows comparisons of different management issues that concern risk, resilience, and sustainability. Figure 6-4 shows how risk is a superset of both resilience and sustainability. It also shows how risk can also be considered a superset of reliability and safety, which are of interest to many engineering disciplines.

6.1.2 Asset Resilience versus Community Resilience

As the name implies, asset resilience is the resilience of a single asset. For our immediate purposes, we consider an asset to be a civil infrastructure asset. For example, a building, a bridge, a mass transit station, and a tunnel are all examples of an asset. DHS (2009) and ASCE (2013) contain a comprehensive list of the types of civil infrastructure assets. Within an

Table 6-3. Risk, Resilience, and Sustainability Comparisons.

Item	Risk	Resilience	Sustainability
Components	• Vulnerability, • Threat, • Consequences	• Robustness, • Resourcefulness, • Recovery, • Redundancy	Varies but can also be traced to vulnerabilities, threats, and consequences.
Management components	Assessment, acceptance, treatment/improvement, monitoring, communication		
Metric properties	• Objective • Subjective • Relative • Absolute (can be monetary, time, or combinations of the same)		
Main emphasis	Cost or cost benefits	Continuity of operations	Long-term consequences Conservation of resources

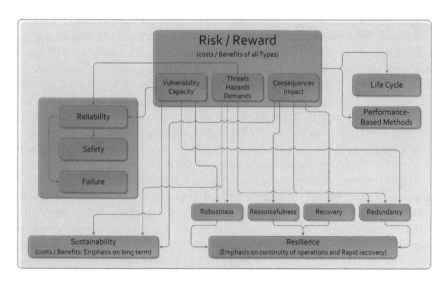

Figure 6-4. Risk, resilience, sustainability interrelationships.
Source: Ettouney (2014).

Table 6-4. Cable Bridge Asset Resilience Example.

Resilience components	Cable bridge parameter (considerations)
R1: Robustness	• Suspension cables • Stiffening girders/trusses
R2: Resourcefulness	• Inspection (special) • Memorandum of Understandings (MOUs) among different organizations.
R3: Recovery	• Approaches (ramps) of roadway to bridge • Training of all kinds
R4: Redundancy	• Approaches (ramps) of roadway to bridge • Suspenders

Source: Alampalli and Ettouney (2016).

asset, different parameters (sometimes referred to as considerations) control asset resilience. These parameters can be categorized as the components of one or more of the 4Rs. Table 6-4 shows a simplified example of a cable bridge's components and categorizations in an asset resilience setting. Note that bridge approaches fit in more than one resilience component (recovery and redundancy). In addition to the categorizations of Table 6-4, a functional diagram, called a network or a graph, needs to be established. This network shows dependencies of, or links among, different parameters. The dependencies are expressed by arrows from the controlling parameter to the dependent parameter. If there is no obvious dependence between two linked parameters, then a simple line connecting the two parameters is used. Figure 6-5 shows a simple network for an asset resilience of a medium-sized building. Capturing important parameters (both operational and physical), as well as their interdependencies, as shown in Table 6-4 and Figure 6-5, is an essential first step for achieving successful asset resilience management. We will discuss the objective processes of networks/graphs in Section 6.2.

We turn our attention now to community resilience. As the name implies, a community is comprised of several assets (nodes) that are interconnected via links that may be assets themselves. The nodes and links constitute a community's network. Community resilience is dependent on the resilience of the network's individual asset components (both nodes and links). Thus, as an essential step of community resilience management, we need to have a greater understanding of the resilience of nodes and links. In addition, community resilience will depend on the topology of the network and how different nodes are linked together. The size of the community is completely subjective. A community could be a

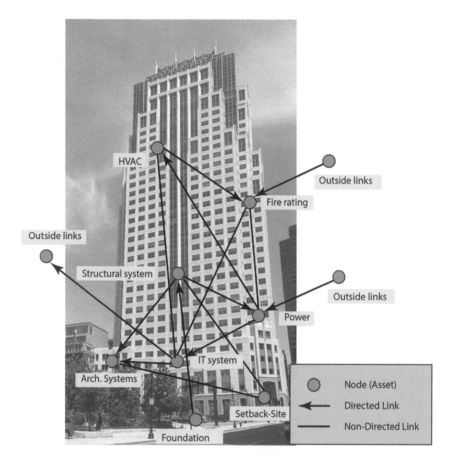

Figure 6-5. Asset resilience links for resilience management (simplified tall building example).
Source: Ettouney (2014).

simple campus comprising a small number of buildings (such as a small hospital or college). A community could be a transportation network, a small town, a region, a whole county, or even a state. Each of the 4Rs of community resilience is a function of all the 4Rs of its nodes and links as well as the topology of the network. Figure 6-6 shows a simple resilience network for a small community.

6.1.3 Essentiality of Network Considerations for an Objective Resilience Management

Our discussions of asset and community resilience so far show that the links among nodes within an asset or community resilience models are an integral part of the model. Moreover, we can state that these links are

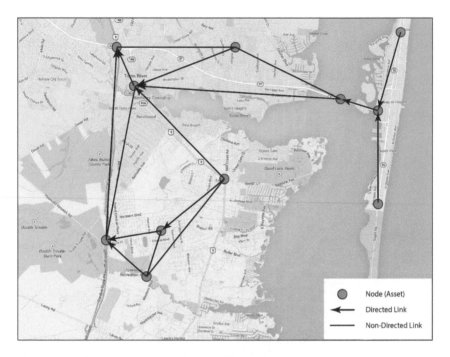

Figure 6-6. Community network for resilience management.
Source: Ettouney (2014).

(1) needed for simulating the operational, decision, and physical processes of assets and communities and can be simulated by either directional or nondirectional links that represent the flow of causes and effects among different nodes (see Figures 6-5 and 6-6).

We then reach an important conclusion: ignoring these links (interactions) while performing any of the resilience management operations can lead to erroneous decisions or computations, which might, in turn, lead to costly operations and the potential for unintended consequences. Most of our examples in the chapter will be based on linked graphs/networks. A summary of the objective underpinnings of the network/graph considerations is explored in Section 6-2.

6.1.4 Objective versus Subjective Resilience Scales

Defining a resilience scale is an essential step in objective resilience processing. Without appropriate resilience scales, decision-making processes required for efficient management will not be optimal. Given that resilience is a special case of risk, however, we can apply well-studied risk scales and descriptors to resilience. As with risk, subjective resilience scales are described as ranks. A three-step ranking system of high, medium, and low is useful for describing resilience (or any of its components). A 5-step rank or 10-step rank can be used. Some authors have utilized this approach to describe resilience in a subjective manner (see Bruneau et al. 2003, Bruneau and Reinhorn 2007). Alampalli and Ettouney (2010) described subjective resilience as a decision-making tool. Subjective resilience scales are all, by definition, relative. They offer simplicity yet cannot be used on large and demanding analytical projects.

Objective resilience scales can be subdivided into two categories: relative or absolute. Again, this is similar to risk (see Kennett et al. 2011a, b, c; Ettouney and Alampalli 2017b). Relative objective resilience (or any of its 4R components) can be described objectively on an analytical scale, from zero to 10 or from zero to 100. The higher the score, the better the situation. A perfect resilience would be at the top of the scale, with a resilience of 100 out of 100. The relative resilience scales were introduced by Kennett et al. (2011a, b, c).

Another objective scale is the absolute resilience scale. It can be measured in units of time or by dollar-time units to reach an acceptable level of postevent performance. An example would be a resilience of 50 days or 35.5 days. The lower number indicates better resilience. The dollar-time resilience scale might read $250 days or $1,025 days. The lower number indicates better resilience. Using only units of time as a resilience metric ignores costs, whereas a dollar-time resilience metric would recognize both costs and time to recovery. The absolute resilience scale was used by Ettouney and Alampalli (2017a b). A comparison of the advantages and disadvantages of resilience scales is shown in Table 6-5.

6.1.5 Multidimensionality of Resilience

The complexities of resilience as we discussed them so far should have implied a multidimensionality aspect of resilience. Specifically, we note that resilience has the following dimensions:

- *Resilience components*: As previously discussed, resilience includes several components, depending on the definition, for example, the 4Rs or the eight components of PPD-8/PPD-21.
- *Resilience management components*: Resilience management [see Ettouney (2014) and Ettouney and Alampalli (2017b)], as we will see in later sections, is composed of five components (assessment, acceptance, treatment, monitoring, and communication).
- *Multidisciplinary components*: Several disciplines/agencies need to coordinate their efforts to provide resilient assets or communities. The efficiency of the coordination depends on accommodating all pertinent disciplines as well as the attributes of each discipline.

Table 6-5. Advantages and Disadvantages of Resilience Scales.

Resilience scale		Advantages	Disadvantages
Subjective	Relative-discrete	Simple	Difficult to use for detailed resilience management components other than assessment.
Objective	Relative-continuous	Simple. Can be used for large-scale assessment and project prioritizations.	Requires more computational resources.
	Absolute	Most useful of resilience scales.	Requires most computational resources. Can be complex at times. In its infancy (as of when this article was written).

Source: Ettouney (2014).

- *Multihazard components*: Individual assets or integrated communities are usually exposed to multitudes of hazards. Depending on the attributes of each hazard, different considerations might be needed to enhance a system's resilience.
- *Cascading components*: In many situations, what starts as a minor event might propagate (cascade) and become a major event. This cascading effect is another resilience dimension that needs to be accounted for, otherwise unintended negative consequences might result.

To achieve the desired accurate objective goals, a modeler of resilience, a decision maker, or a stakeholder needs to be specific/clear as to which dimension is under consideration. Mixing different resilience dimensions out of the aforementioned five dimensions will usually produce more meaningful, realistic, and accurate results. Figure 6-7 shows a process for accommodating resilience multidimensionality in modeling, decision-making, or general resilience management activities.

The next section will present some of the objective processes that can be used to handle different resilience dimensions, and the remainder of the sections will explore how to apply these objective processes to each of the resilience dimensions.

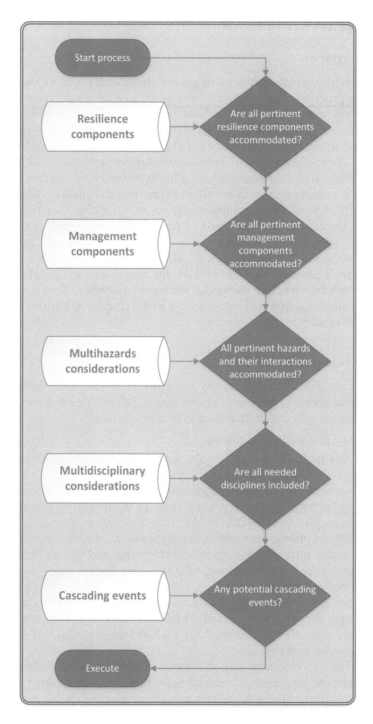

Figure 6-7. Process for accommodating resilience multidimensionaliy.

6.2 OBJECTIVE PROCESSES

6.2.1 Overview

As previously stated , the main goal of this chapter is to explore some objective processes that might be considered to address different aspects of resilience management. Note that resilience management is a complex field with several components such as preparedness, robustness, recovery, and so on, whereas the management of each of these components has itself several components such as assessment, acceptance, mitigation, monitoring, and communication. The confluence of resilience and resilience management will result in numerous dissimilar components with each of these components requiring its own objective processing method. Because of this, we need to consider several types of objective methodologies and classes.

This section will first overview the different resilience metrics that can be objectively utilized. We then overview the advantages and disadvantages of two popular objective methods. We finally overview four promising objective methods that will be used later in the chapter to address various aspects of resilience management.

6.2.2 Resilience Metrics

Our objective treatment of any issue needs to start with identifying appropriate metrics for describing the issues. For resilience, there are two metric categories: subjective and objective as follows:

- Subjective resilience metrics:
  - Abstract metrics that describe the totality of resilience, for example, [High, Medium, Low] or [High, Low].
  - More descriptive metrics such as describing the state of continued operations or the recovery time in a subjective manner. For example, to subjectively describe continuity of operations: no interruptions, minimal interruptions, moderate interruptions, severe interruptions, complete stoppage, and so on. To describe recovery time: [Very long, Medium, Short, Minimal]
- Objective resilience metrics:
  - Several objective metrics are available that can describe continued operations such as a set of actual times of operation interruptions, for example 0.0, 5.0 min, 0.5 h, 5.0 h, 0.5 days, 1.0 day, 1.0 week, 5.0 weeks, and so on..
  - Time to recovery can also be used as an objective metric in a similar manner as continued operations, aforementioned.
  - Monetary metric is another important objective metric that can be used to evaluate resilience. In such a situation, as discussed in

Section 6.1.1.3 and by Ettouney and Alampalli (2011b), Ettouney (2014), and Ettouney and Alampalli (2017b), resilience becomes another manifestation of risk.

6.2.3 Popular Objective Methods

There are two popular objective methods that are used to evaluate resilience: specifically, analytical methods and the weighted averages method. Obviously, there are reasons for their popularity. Some drawbacks also exist. We briefly discuss each of these approaches next.

6.2.3.1 Analytical Methods. Evaluating risk, or resilience, using analytical methods is enticing because it can utilize simple formulas that are amenable to mathematical manipulations and can produce finite expressions for resilience; see Vose (2009). The main attractions of this approach are its elegant formality (both input and output forms) and the ease of tracing causes and effects. Thus, it can provide powerful research tools.

One of the disadvantages of analytical handling of a complex subject such as infrastructure resilience is that resilience is a multifaceted and complex issue that includes numerous variables that contribute to the problem in a highly nonlinear manner. Expressing such a complex subject via analytical means will require numerous simplifying assumptions. Such simplifying assumptions will lead to a result with less accuracy. Because of this trade-off between the needed simplifying assumptions and the accuracy of results, we will not consider analytical processes in this chapter.

6.2.3.2 Weighted Averages Method. The weighted averages approach has been extensively used for evaluating the resilience of different types of infrastructure; see Kennett et al. (2011a, b). They offered a resilience assessment procedure for buildings, mass transit stations, and tunnels, respectively. Chavel and Yadlosky (2011) addressed the resilience of bridge design, whereas Mertz (2012) described bridge physical redundancy (structural, load path, and internal). Hughes and Healy (2014) published a resilience methodology for bridges in New Zealand. The main advantage of the weighted averages method is its simplicity. The main disadvantage is that it is not easy to account for the interactions among variables, especially in large models. Ettouney and Alampalli (2017b) showed that the method can produce inaccurate results when compared with other methods that consider interactions among variables. Therefore, we will not use the weighted averages method in the remainder of this chapter.

6.2.3.3 This Section. In the remainder of this section, we briefly discuss four types of objective and semiobjective processes that we will be using in the different examples throughout the chapter. Because of space limitation, we present only the major aspects of each process. For a more in-depth understanding of these processes, we will cite several references for more detailed explanations.

6.2.4 Networks and Their General Components

A detailed study and an accurate analysis of the interrelationship of infrastructure risk or resilience require accommodating several attributes noted in the following. An improper consideration of these attributes may lead to incomplete and/or inaccurate results, which, in turn, can lead to decisions with potential negative consequences. Attributes of interest include the following:

- Realistic and accurate modeling of linkages/interactions among variables.
- Accommodation of uncertainties associated with different variables.
- Accurate modeling of observations in a temporal scale (snapshot or a finite time period).
- Accommodation of both objective and subjective variables and pertinent combinations.
- Seamless accommodation of decisions under uncertainty, while accounting for all the aforementioned attributes.
- Ease of handling large and/or complex models.
- Efficient use of computational resources.

As previously discussed, neither analytical solutions nor the weighted averages method can accommodate the aforementioned requirements. This limitation can lead to inaccurate results. However, a network-based method is capable of addressing all the requirements. Specifically, graph networks (GNs) and probabilistic graph networks (PGNs) can fulfill these requirements when used for resilience management.

GN and PGN methods have been extensively used in the recent past in computer science and medicine, as well as several other fields; see Deo (1974) for GN-based methods and Koller and Friedman (2009) for PGN applications. Because there are several commonalities between GNs and PGNs, this section will introduce some of the important commonalities. Section 6.2.4 will discuss some of the specifics of GNs, and Section 6.2.5 will discuss the same for PGNs.

In general, a graph network, $\mathcal{G}$, is defined as

$$\mathcal{G} = \mathcal{G}(\mathcal{X}, \mathcal{L}) \tag{6-2}$$

In this equation, $\mathcal{X}$ represents a finite set of nodes (sometimes referred to as "vertices") and $\mathcal{L}$ represents a finite set of links (sometimes referred to as "edges" or "connectors") between the nodes. In general, the nodes represent variables $X_i \in \mathcal{X}$. Within $\mathcal{G}$, an ordered pair of nodes, X_i and X_j, are connected by a link $\ell(X_i, X_j)$ such that $\ell(X_i, X_j) \in \mathcal{L}$.

Two types of links, $\ell(X_i, X_j)$, are available in a network: nondirected links and directed links. The nature and method of processing each type of link differ in GNs and PGNs. We discuss each in Sections 6.2.4 and 6.2.5, respectively.

6.2.5 Graph Networks

A graph network (GN), $\mathcal{G}(\mathcal{X}, \mathcal{L})$, can be used to provide objective measures of many resilience aspects. This can be done by utilizing the properties of the nodes, links, and/or the whole network. There is a large body of work that studies GNs and its numerous properties; see Deo (1974), Joyner et al. (2012), and Newman (2010). Our main interest in this chapter is to show how some of the GN properties can be used to address some of the needs of resilience management. The GN properties that we will be using later in this chapter are summarized next. For more details on any of these properties, see Deo (1974) or Newman (2010).

6.2.6 Some Important Graph Networks Properties

Path length, $Pl(a, b)$, is the distance between two nodes a and b. Formally,

$$Pl(a,b) = \sum_{\ell(X_i, X_j)} Wt\left[\ell(X_i, X_j)\right] \ell(X_i, X_j) \tag{6-3}$$

with links $\ell(X_i, X_j)$ providing an uninterrupted connection between nodes a and b.

$$Pl(a,b) = \infty, \text{ if there are no edges connecting } a \text{ and } b \tag{6-4}$$

$$Pl(a,b) = 0, \quad \text{if } a = b \tag{6-5}$$

$$\ell(X_i, X_j) = 1, \quad \text{if } i \text{ and } j \text{ are adjacent} \tag{6-6}$$

The weight of the link $\ell(X_i, X_j)$ is $Wt[\ell(X_i, X_j)]$. It can be used to indicate the strength of the link. In some applications, such as transportation, it can be used to indicate cost or impedance. If all links have similar weight, then $Wt[\ell(X_i, X_j)] = 1$. Path length, $Pl(a, b)$, can be used in several instances; for example, it can indicate how close two organizations are to each other in a multiorganizational network.

Graph length, Lg, is the total path lengths within a graph. Formally,

$$Lg = \sum_{\mathcal{L}} Wt[\ell(X_i, X_j)] \, \ell(X_i, X_j) \qquad (6\text{-}7)$$

Graph length, Lg, is used in transportation networks to compute the total length of roads in a network. It can also be used to estimate the efficiency of communication among stakeholders in a GN.

Geodesic distance, $gd(a, b)$, is defined as the shortest path between a and b. Formally,

$$gd(a,b) = \mathrm{MIN}(Pl(a,b)) \qquad (6\text{-}8)$$

One possible use of geodesic distance, $gd(a, b)$, can be used to find an optimal route for vehicles during an emergency.

Degree of nodes $Dg(i)$:

$$Dg(i) = \sum_{j} \ell(i, j) \qquad (6\text{-}9)$$

with j spanning all edges that connect to node i. The degree of a node is an indication of how well-connected the node is to the rest of the GN.

In degree $IDg(i)$:

$$IDg(i) = \sum_{j} \ell(i, j) \qquad (6\text{-}10)$$

with j spanning all inward directed edges that connect to node i in a directed graph. The in degree of a node is an indication of the receiving capabilities of the node within the GN.

Out degree $ODg(i)$:

$$ODg(i) = \sum_{j} \ell(i, j) \qquad (6\text{-}11)$$

with j spanning all outward edges that connect to node i in a directed graph. The out degree of a node is an indication of the nodal function as a source within the GN.

Centrality: Several metrics are available for the centrality of a node in a graph. We will look at only two of these metrics: degree centrality and closeness centrality. For other types of centrality, see Deo (1974).

Nodal degree centrality: This is the same as $Dg(i)$; see Equation (6-9). The higher the nodal degree centrality is, the more centralized the node within the GN is. Note that this measure is a localized measure that does not account for the location of the node within the whole network structure.

Closeness centrality, $Cl(i)$: The closeness centrality of a node is defined as the sum of all geodesic distances between that node and all other nodes in the graph. Formally,

$$Cl(i) = \sum_{j} gd(i, j) \qquad (6\text{-}12)$$

The closeness centrality of a node measures how close is the node, on average, to all other nodes in the network. For example, the location of an emergency center such as a hospital should have a low closeness centrality within the community.

Diameter of a graph, δ: The diameter of a graph is a measure of the extent of the graph as a whole. It is defined as the longest geodesic distance between any two nodes in a graph. Formally, it is expressed as follows:

$$\delta = \underset{(X_i, X_j)}{\mathrm{MAX}}[gd(X_i, X_j)] \qquad (6\text{-}13)$$

Pi index, π: This index is a measure of the density and complexity of the graph. A higher π indicates a well-developed and dense graph. Formally,

$$\pi = \frac{Lg}{\delta} \qquad (6\text{-}14)$$

The π index can be used to evaluate the properties of a transportation network. It can also be used to evaluate the efficiency and complexity of multidisciplinary networks.

6.2.7 Probabilistic Graph Networks

What differentiates PGNs from GNs is that the nodal variables in PGNs represent random variables. In addition, decision models of PGNs will have more types of nodal variables such as decision and utility variables. In addition, dynamic PGNs will have time link nodes (TLNs). Links in PGNs are also different from those in GNs. In PGNs, directed links are described by using a conditional probabilities table (CPT) that contains conditional probabilities of the states of the child (end) node given the states of the parent (source) node. Nondirectional links are described by using potentials that contain the affinities among the states of the different nodes within a group of nodes (known as "cliques") that are connected by the nondirectional links. A clique may contain two or more nodes. The CPTs and/or potentials within a PGN can then be solved to produce the historical marginal probabilities of all random variables. Of course, the true power of a GN lies in its capability of inference or introducing evidence/observation of one or more variables within the network and calculating the corresponding probabilities of other variables given those observations.

As an objective tool, PGNs are eminently suitable for different applications of resilience management. Some of the advantages of PGN use are as follows:

- They are probabilistically based. They can handle the inherent uncertainties of resilience and all of its components.
- Links and dependencies of different parameters are built-in.
- Adding or removing nodes or links from a PGN model can be done easily.
- Forecasting and life-cycle analysis, see Section 6.2.5.5, and decision-making, see Section 6.2.5.4, can be integrated easily with PGNs.

For a more in-depth coverage of PGNs, see Fenton and Neil (2013), Neapolitan (2004), or Koller and Friedman (2009). Several different types of PGNs are used as examples in this chapter. We will introduce them briefly next.

6.2.7.1 Bayesian Networks.   The nodal variables in BNs are all random variables. These random variables can be discrete or continuous. Moreover, the interrelationships among the variables are described by using CPTs. For a detailed description of BNs, see Fenton and Neil (2013) or Koller and Friedman (2009).

6.2.7.2 Markov Networks.   Similar to BNs, the variables in MNs (sometimes referred to as Markov random fields) are also random variables. The main difference between BNs and MNs lies in the fact that the links between a subset of variables are nondirectional. Linked variables will mutually affect one another through a potential matrix, instead of the CPTs of BNs. Fenton and Neil (2013) have given a detailed description of MNs, their properties, and solution methods.

6.2.7.3 Chain Graph.   In many practical situations, a need may arise to use a mix of directional and nondirectional links among the variables of a PGN. The resulting PGN is called a chain graph (CG). The name CG derives from the fact that the model is subdivided in a chain of a superset of variables that are connected by a chain of a superset of CPTs. Details of CGs and their solution processes can be found in Koller and Friedman (2009).

6.2.7.4 Influence Diagrams and Decision-Making. Influence diagrams are a special kind of PGN that supports decision-making. To provide this capability, IDs include two additional types of variables, in addition to random variables. A decision variable and a utility variable need to be included in an ID. Note that an ID can include several decision variables. Similarly, an ID can include several utility variables. The links in an ID can be any needed mix of directional and nondirectional links. The objective of the ID model is to find a decision, or a set of decisions, which will result in an optimal utility, given the random variable set of the model. (Note that in an ID problem, decision states of a decision variable are the available *policies* to the decision maker. Also, a decision node should be considered as a policy node. A *strategy* is the path from decision nodes, which are decision-dependent, to the utility node, through other variables in the model. An optimal *strategy* is the strategy that will result in an optimal utility. In our treatment of ID models, decisions and policies will be used interchangeably.)

The steps to building an ID-based decision model are summarized as follows:

1. Define controlling issues and model as nodes in the ID.
2. Define available decisions (policies) to decision makers.
3. Establish utilities for different decisions.
4. Establish links among different variables in the model.
5. Establish conditional probabilities and/or potentials for links.
6. Describe any observed variable.
7. Solve the probabilistic influence diagram (ID) model to find the decision (policy)/strategy combination that would produce an optimal utility, or a set of utilities, given the observations in Step 6.

A detailed description of ID processes can be found in Neapolitan (2004).

6.2.7.5 Decision-Making, Policy, and Strategy. Any discussion of decision-making should lead us to the subject of policy and strategy. Unfortunately, there is some confusion in using the terms policy and strategy (see Koller and Friedman 2009, Powell 2011). Surprisingly, such confusion extends to even some objective processes. Because resilience-related decision-making will involve choosing policies and strategies, we need to clarify the difference between the two. For our purposes, we will use policy to indicate a decision taken among several available decisions (policies). As such, we will be using policy and decision interchangeably throughout this chapter. A strategy is a set of actions that a particular policy (decision) needs to take to achieve a particular goal. Several strategies exist for a given policy. However, only one strategy of these will be an optimal strategy, which is the one that produces an optimal result/goal. Following

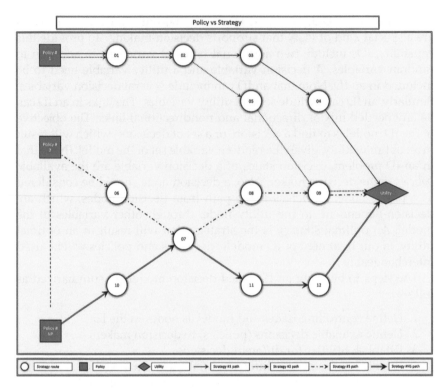

Figure 6-8. Policy versus strategy in a decision-based network, for example influence diagram.

these definitions of policy, strategy, and optimal strategy, we note further the following:

- Single policy might produce multiple strategies.
- Single strategy might be available to single or multiple policies.
- Multiple policies might produce multiple strategies.

Figure 6-8 shows a decision-based network, for example, an influence diagram (ID). The figure shows *NP* policies (decisions) with *NS* possible strategies that span the network. All strategies lead to the desired goal (utility). At the utility node of the network, the optimal choice of a policy strategy can be chosen.

6.2.8 Dynamic Probabilistic Graph Network

Almost all properties of civil infrastructure are never static. They always change as time progresses. They include vulnerabilities, hazards, and consequences. As such, all resilience components that relate to civil infrastructure are time-dependent. This time dependency requires objective methods that can account for temporal changes. An extension of all PGN types of this section can be developed to accommodate the needed time marching. For more information about the dynamic probabilistic graph network (DPGN), see Koller and Friedman (2009), Neapolitan (2004), or Ettouney and Alampalli (2017b). To account for time marching, the DPGN adds three dynamic properties to static PGNs:

1. At each time snapshot, a PGN topology is used (called time slice).
2. Time slice includes new types of nodes within the PGN: time link nodes (TLNs). These TLNs are usually present in all time slices.
3. Transfer matrix is defined for each TLN. The transfer matrix defines the changes of probabilities of the TLNs from one time slice to the next time slice.

Using these three dynamic properties, the time-marching solution of the DPGN is obtained as follows:

1. Solve the initial time slice, including any nodal observations.
2. Extrapolate the just-computed probabilities of all TLNs to the next time slice using the pertinent transfer matrices.
3. Extrapolated probabilities of the TLNs in the new time slices are then used as soft observations, see Fenton and Neil (2013), to solve the new time slice of the DPGN.
4. Repeat Steps 1 through 3 for as many time slices as desired.

The aforementioned process can be used for any type of PGN to forecast resilience and any of its components and then to optimize future decisions (policies/strategies). The DPGN can also investigate past performances by altering the transfer matrices and performing a backward time marching.

6.2.9 Game Theory

6.2.9.1 Overview. Game theory is a branch of mathematics and economics that has been successfully used in solving many problems in many applications; see Fudenberg (1991), Gibbons (1992), or Prisner (2014). We will briefly discuss the basic components of the game theory in this section with emphasis on particular applications that we will use later to objectively study some aspects of resilience management.

6.2.9.2 Components of Games. Each game will have these components:

- Game is usually played by at least two players; many players can participate in the game. The players can be individual people, companies or organizations, countries, or a mix of any of these entities.
- Each player in a game will need a defined set of payoffs that might be gained (or lost) depending on the type of game and how the

game is set up. The payoffs for each player will depend on the set of strategies that are available to that player. Payoffs might be objective (e.g., monetary), binary (e.g., yes/no, accept/reject, or win/lose), or subjective (e.g., high, medium, low). The combinations of these payoffs would constitute a payoff table that can be two dimensional for two-player games or multidimensional for multiplayer games.

- Each game will need a defined set of rules.

6.2.9.3 Plethora of Games and How Games are Solved. Several classes of games are available, see Prisner (2014). These include, but are not limited to,

- Zero, or constant, sum games.
- Cooperative games.
- Complete versus incomplete information games.
- In-series versus in-parallel games.
- Potential for random moves in a game.

Within each of the aforementioned classes, there are numerous types of games. Some well-known games are prisoner's dilemma, screening game, signaling game, and public indifference game (sometimes called public apathy game).

Several solution methods exist for games that depend on different attributes of each game. Among the solution methods are dominant strategy equilibrium and Nash equilibrium.

We will explore applications of the public indifference game and its solution by both methods later in this chapter. Many other game solution methods include Bayesian equilibrium and weak form Bayesian equilibrium. However, a complete discussion of game solution methods is beyond the scope of this chapter; see Fudenberg (1991), Osborne and Rubinstein (1994), or Spaniel (2015) for a more comprehensive discussion of the subject.

6.2.9.4 Public Indifference Game and Its Solution Methods. *The game of public indifference*: Investigating resilience (or risk) issues that relate to assets and communities will invariably involve the public as one of the players. The term "public" in the current context could mean dwellers of a particular building, employees of a business, residences of a community, or even a county or a state. Of interest is the public sentiment, which can be redefined as a payoff in the game. The other players in the game might be business owners, asset or community mangers, or public relations organizations; in short, any entity that needs to deal with the "public." The payoffs of that entity can be costs, messaging, see Section 6.3.6.1, or preparedness strategies, see Section 6.7.4.

Note that the public indifference game is not a one-time game, rather it is a repeated game. As such, the solution of the game is a long-term solution. The public sentiment in this game (payoff) is measured by a binary outcome (accept/reject).

The objective of the game is that the organization that is encountering the public should find the optimal strategy or strategy mix that will maximize the possibility of public acceptance to its strategy, or strategy mix.

Solution: Dominant Strategy Equilibrium: A *dominant strategy* for a player is the strategy that produces the best outcome no matter what any other player, or players, do. A game will result in a *dominant strategy equilibrium* when *all* players practice a dominant strategy; see Sections 6.3.6.1.2 and 6.7.4.2. This is perhaps the easiest solution of the game. Unfortunately, such a coincidence does not always happen, and, therefore, there is a need for a more involved solution.

Solution: Mixed strategies and Nash equilibrium: The optimal strategy, or mixed strategy equilibrium of the organization, will occur when this mix will make the public indifferent to all the strategies of the organization. Objectively, this will happen when the expected utilities, as measured by public payoffs, of all of the organization's available strategies are equal; see Osborne and Rubinstein (1994).

Suppose we assume that the number of strategies available to an organization and the public is N_s. We also assume that the frequency (probability) of the organization using the ith strategy is p_i, with $i = 1, 2, \dots$ $(N_s - 1)$ and $p_{N_s} = 1.0 - \sum_{i=1}^{i=(N_s-1)} p_i$. The expected utility $EU[S_i]$ for the response of the public to the ith strategy of the organization will be

$$EU[S_i] = \sum_{j=1}^{j=N_s} (p_j P_{ji}) \qquad (6\text{-}15)$$

with P_{ji} being the public payoff value for the jth strategy of the organization and the ith strategy of the public. The conditions of making the public indifferent to all the strategies of the organization can be satisfied by enforcing

$$EU[S_i] = EU[S_{i+1}] \qquad (6\text{-}16)$$

for $(N_s - 1)$ linear equation with $(N_s - 1)$ unknowns. Solving these equations will produce desired probabilities, p_i, with $i = 1, 2, \dots (N_s - 1)$, and finding the final probability P_{N_s} is trivial. The organization now can attain the optimal public reaction by mixing its strategies using frequencies according to p_i, $i = 1, 2, \dots N_s$.

Note that we assumed an equal number of strategies that are available to both the organization and the public. This assumption can be released. In such a case, the system of equations (6-16) will not be square, and different types of solutions are needed. A system that is based on an optimization technique such as singular valued decomposition is used in practice to solve this type of equation. However, this situation is beyond the scope of this chapter.

Solving equations (6-16) for $N_s = 2$ can be done graphically and/or numerically; see Section 3.6.1.3. For $N_s > 2$, a graphic solution is difficult; however, numerical solutions can be used. An example for $N_s = 4$ is offered in Section 7.4.3.

6.2.10 The Ostrich Paradox

6.2.10.1 Overview. A study by the Multihazard Mitigation Council (MMC 2005) showed that a savings of $4.0 is incurred for every $1.0 spent on the mitigation of/preparedness for natural hazards. However, Meyer and Kunreuther (2017) observed that there seems to be a public bias toward preparedness efforts, in spite of those reporting high benefit-to-cost ratios. On studying this issue further, Meyer and Kunreuther (2017) concluded that there are six preparedness impedance biases (PIBs). They called this phenomenon "the Ostrich Paradox." We summarize the six PIBs in this section, even though they were not presented by Meyer and Kunreuther (2017) as a formal objective theory, because they embody several important principles regarding preparedness. We will be using PIBs, as well as some of the principles and solutions of the Ostrich Paradox, as a basis for some objective models later.

6.2.10.2 The Biases against Preparedness. Figure 6-9 illustrates the six PIBs and their subimpedances. We can summarize them as follows:

1. *Inertia bias:* Many individuals and organizations find it easier to keeping status quo. Of course, by doing this, they fail to prepare for the potential of the changing demands for future hazardous events.
2. *Optimism bias:* This results from an unfounded optimism by the public that future hazards might not affect them, or if they do affect them, the damage will not be as severe. There are some sub-biases of optimism, including the following:
   a. *Availability:* Extrapolating personal, or organizational, own risk experiences to other risk situations that might not be similar in nature.
   b. *Anchoring bias:* Conflating, or anchoring, long-term risks with short-term risks.

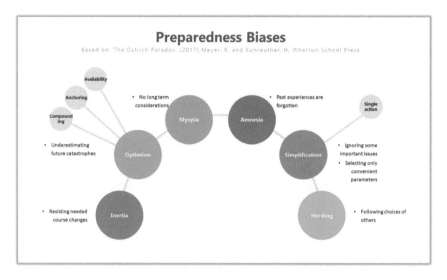

Figure 6-9. Different biases that impede preparedness.

    c. *Compounding bias:* Ignoring the compounding effects of high-risk, low-probability events such as climate change or terrorist attack hazards.
3. *Myopia bias:* Source of this bias is the failure to forecast events and their potential damaging consequences correctly.
4. *Amnesia bias:* Failure to apply the lessons of past damaging events to prepare for the future.
5. *Simplification bias:* This bias results from tendencies to oversimplify complex risks, thus underestimating potential damaging consequences. A resulting sub-bias of simplification is Single action bias: Oversimplification of complex risk potential might lead to considering only a subset of the risk, which might lead to a single preparedness action, instead of preparedness on multiple fronts.
6. *Herding bias:* Tendency of individuals or organizations to follow the leads of others, rather than charting their own courses.

A comprehensive discussion of the PIBs can be found in Meyer and Kunreuther (2017).

6.2.10.3 Concluding Remarks.   The Ostrich Paradox promises to be the basis for many objective studies and developments in the fields of resilience, risk, and climate change. With appropriate objective modeling, including those in this chapter (Sections 3.2.4.2, 3.2.4.3, and 7.3.2), resilience management stakeholders can provide suitable preparedness efforts that should enhance the resilience of their entities.

6.3 COMPONENTS OF RESILIENCE MANAGEMENT

6.3.1 Overview

After introducing resilience and its components in Section 1.1 and the dimensions of resilience in Section 1.4, we turn our attention to one of those dimensions: resilience management and how stakeholders can manage the resilience of their assets or communities. It is usual that the first step in the management effort is to *monitor (or inspect)* (We use the term "monitor" here in a general sense. It includes visual inspection, estimation of attributes using "expert knowledge," mechanical and electrical sensors, as well as any available subjective or objective means that results in a reasonable estimation.) the states of different resilience parameters to establish needed information for *assessing* the state of resilience amid a natural or man-made hazard or threat. Second, there is a need to decide whether to *accept* this assessment. If it is decided that the assessed resilience level is not acceptable, a resilience *improvement* program should be initiated to bring the assessed resilience up to an acceptable level. Unfortunately, the passage of time has a tendency to reduce physical capacity (aging effects) and/or operational efficiency (budget cuts, for example) while demands from infrastructure tend to increase (e.g., traffic, intensity of storms, climate change). These universal truths will have the inescapable result of resilience reduction as time goes on. This will necessitate the need for resilience *monitoring* (*ReMo*) to ensure that resilience remains within an acceptable level. Finally, all of those activities need to be *communicated* among all stakeholders, including the legislators and the public, to maintain public confidence and preparedness as well as to avail an adequate level of project funding.

All of the aforementioned five components (assessment, acceptance, improvement, monitoring, and communication) constitute an essential framework for a complete and successful resilience management, as shown in Figure 6-10. The figure also reveals the complex interrelationships among the five resilience management components. Figure 6-10 illustrates that resilience management should be a continuous process. Many of the parameters that affect asset or community resilience vary with time, and, hence, a monitoring of these parameters needs to be done on a regular basis. Then, when needed, reassessment, acceptance, and improvements need to be done. Pertinent communication among all stakeholders must be initiated while all of these activities are underway. Additional notes are given in Figure 6-10:

- Communication is at the center of all management phases. Decision makers should always be in communication with all phases of the process.
- The processes of handling resilience should always start with the monitoring phase. This is due to the fact that monitoring si the source

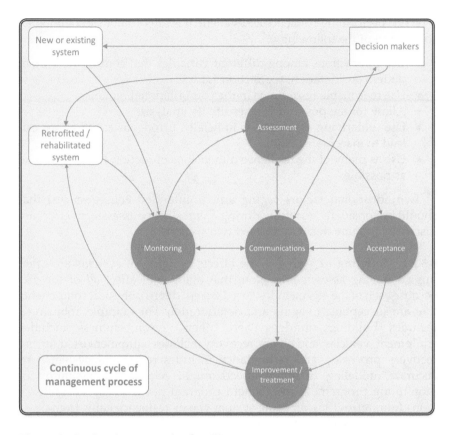

Figure 6-10. Continuous cycle of resilience management.

    of all assumptions and/or information that control the assessment
process.
- Cycle always starts and ends at the decision maker level. It should
continue on a life cycle basis until the system is decommissioned.

6.3.2 Resilience Assessment

    Assessment component in resilience management involves evaluating
the resilience level of either the asset or the community of interest. Because
resilience is a special representation of risk, it is natural to utilize risk
assessment methods when trying to assess resilience, if needed. Some of
the methods for assessing risk are based on subjective processes (Gutteling
and Wiegman 1996), whereas others are based on objective processes
(Kennett et al. 2011a, b, c). Detailed relative and absolute risk assessment
methods are also available; see Ettouney and Alampalli (2011a, b). Finally,
probabilistic or deterministic risk assessment methods can be found in
Fenton and Neil (2013) and Ettouney and Alampalli (2017a, b).

While performing resilience assessment, it is important to accommodate some or all of the following:

- Use interactions among different variables that control the system under consideration; see Section 1.2.1.
- Use reasonable resolutions in the computational model.
- Allow for the possibility of sensitivity analysis.
- Use underlying assumptions judicially. Erroneous assumptions can lead to inaccurate results.
- Use as many of the resilience dimensions of Section 1.4 in the model as possible.

Two additional factors (aging and unintended consequences) that should be considered carefully during any resilience assessment effort are discussed in more detail in the next two sections.

6.3.2.1 Modeling of Aging.   The effects of aging of systems are often missed during assessments. Note that aging will affect *all* of the 4Rs because each of the 4Rs includes, to a different degree, physical components that are susceptible to aging and deterioration. For example, robustness includes buildings, bridges, and others, resourcefulness includes equipment, vehicles, and others, recovery includes equipment used during recovery processes, and redundancy includes redundant systems. An accurate modeling of aging effects might require an institution of monitoring program. A cost/benefit program needs to assess resilience gains from retrofitting an aging system versus resilience gains from other improvements, including operational and resource improvements.

6.3.2.2 Assessing Unintended Consequences.   One of the most desired goals of adequate resilience management is to try to avoid unintended consequences. Some of the concerns that can produce unintended consequences are the following: ignoring interactions within an asset or community, cascading events, the efficiency of a multidisciplinary network, or multihazard interactions. Again, accounting for as many resilience dimensions of Section 1.4 might help in reducing those unintended consequences.

6.3.2.3 Assessing Civil Infrastructure as the Center of a Global Network of Risks.   Section 1.2.1 argued that successful considerations of objective resilience need to account for the nature of network of its controlling variables. Network considerations for objective resilience are even more vital when civil infrastructure are involved. This applies when we consider both single asset resilience and the resilience of communities that include several infrastructure. To illustrate the centrality of civil infrastructure within a network of global risk, we consider how the World Economic Forum (WEF 2017) presented a global risk network and how those risks are linked. To formally address a global risk network, we model it using a GN with the global risk variables (nodes) that are similar to the World Economic Forum (WEF 2017) variables, as shown in Table 6-6. The topology of the network is shown in Figure 6-11. The links in this GN are similar to the links of the World Economic Forum (WEF 2017). The nodes of the network of Table 6-6 and Figure 6-11 are categorized following the categorization of the World Economic Forum (WEF 2017) with the addition of a new category that contains global risks that involve civil infrastructure, such as node #2 (natural disasters). Out of a total of 30 global risks in the network, there are 8 global risks that relate directly to civil infrastructure.

The GN enables us to objectively study the properties of the global risk network. Perhaps the most obvious property of this GN is the dense topology relating to global risks, as shown by a computed network diameter, δ, of 3. This indicates that the global risks are well interconnected. The global risk with the highest degree centrality, $Dg(i)$, is *Failure of regional governance*, with a degree of 23, which is not too surprising, because such a failure can easily lead quickly to many other risks. The least degree centrality is for *Data Fraud of Theft* with a degree centrality of 6; again, not too surprising, because this risk, although important, is not, relatively speaking, as damaging as other global risks.

Turning our attention to civil infrastructure nodes and computing how central they are to other global risk hazards, we develop Table 6-7. The table shows the degree centrality and the closeness centrality for each of the civil infrastructure hazards. Civil infrastructure nodes are well centered and closely related to all other global risks in the network. This means that the considerations of the links between global risks and civil infrastructure are essential; otherwise, an assessment of resilience (or risk) as resulting from other global risks might yield incorrect results.

The graph of Figure 6-11 can be employed to evaluate how close each of the civil infrastructure nodes is to each other (measured by the number of links of the shortest path between two nodes). Table 6-8 shows the geodesic distance, $gd(a, b)$, among the different infrastructure nodes. These nodes are fairly close to one another. We note that the matrix in Table 6-8 is a form of a multihazard matrix that has not been investigated as of the writing of this manuscript. It reflects the interactions among different civil infrastructure risks. A closer inspection of Table 6-8 shows that some risks are more linked together than others. For example, *Terrorist* attacks are closely linked to *Man-made environmental disasters* and *Failure of regional* governance than to *Natural disasters.*

We close this section by noting that the values of those interactions are in the range of 0 to 2, which is a fairly small range. This is because of the coarseness of the network itself. For an in-depth analysis of the interactions, a higher resolution network needs to be developed. However, this subject is beyond the scope of this chapter.

Table 6-6. Key for Node Numbers of the Global Risk Network.

ID	Node name
1	Biodiversity loss
2	Natural disasters
3	Man-made environmental disasters
4	Extreme weather events
5	Spread of infectious diseases
6	Food crises
7	Failure of CC mitigation and adaptation
8	Water crises
9	Failure of critical infrastructure
10	Failure of regional governance
11	Failure of urban planning
12	State collapse or crises
13	Failure of national governance
14	Energy price shock
15	Large-scale involuntary migration
16	Weapons of mass destruction
17	Interstate conflict
18	Profound social instability
19	Illicit trade
20	Unmanageable inflation
21	Critical information infrastructure breakdown
22	Terrorist attacks
23	Unemployment or underemployment
24	Cyber attacks
25	Adverse consequences of technical advances
26	Fiscal crises
27	Asset bubbles
28	Data fraud or theft
29	Failure of financial mechanism or institution
30	Deflation

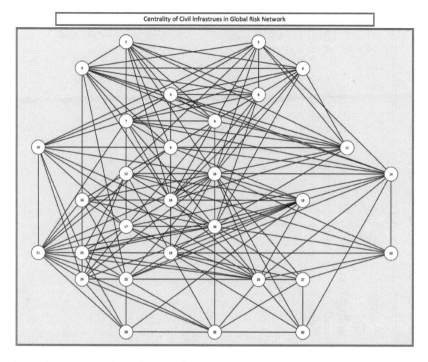

Figure 6-11. Centrality of civil infrastructure within the global risk network. See Table 6-6 for names of different nodes.

Table 6-7. Centrality Measure of Civil Infrastructure to Other Global Risks.

Node #	Node name	Degree centrality, $Dg(i)$	Closeness centrality, $Cl(i)$
2	Natural disasters	12	46
3	Man-made environmental disasters	15	43
4	Extreme weather events	11	51
7	Failure of CC mitigation and adaptation	15	43
8	Water crises	13	46
10	Failure of regional governance	12	46
11	Failure of urban planning	12	47
22	Terrorist attacks	15	43

Table 6-8. Geodesic Distance, $gd(a, b)$ Matrix between Civil Infrastructure Nodes.

	Natural disasters	Man-made environmental disasters	Extreme weather events	Failure of CC mitigation and adaptation	Water crises	Failure of regional governance	Failure of urban planning	Terrorist attacks
Natural disasters	0	1	1	1	1	1	1	2
Man-made environmental disasters	1	0	1	1	1	2	1	1
Extreme weather events	1	1	0	1	1	1	1	2
Failure of CC mitigation and adaptation	1	1	1	0	1	1	1	2
Water crises	1	1	1	1	0	2	1	2
Failure of regional governance	1	2	1	1	2	0	1	1
Failure of urban planning	1	1	1	1	1	1	0	2
Terrorist attacks	2	1	2	2	2	1	2	0

6.3.2.4 Preparedness: An Ostrich Objective Viewpoint

6.3.2.4.1 Ostrich Paradox PGN Model. BIPs describe the effects of those biases that hinder preparedness for disasters. It is desirable to describe these effects in an objective model that can be of help in many objective resilience and risk modeling efforts; see Section 7.3.2. On further reflection, one possible model would be a logistic regression model. We recognize this model as an inverse of the well-known naïve Bayesian model; see Fenton and Neil (2013) or Koller and Friedman (2009). The simplest objective BN model relates the six BIPs, see Meyer and Kunreuther (2017), and a variable that represents the preparedness level. The seven-variable logistic regression BN model is shown in Figure 6-12 and the descriptions of its variables are summarized in Table 6-9.

Note that the logistic regression model is not computationally efficient. The CPT of the preparedness variable would be fairly large because the variable has six parents; see Fenton and Neil (2013). In addition, preparedness for hazards is dependent on several other parameters that are not included in the model, which makes this model incomplete. We will offer computational improvements to the PIBs' model of Figure 6-12 in Section 3.2.4.2 and then offer a BN preparedness computational model

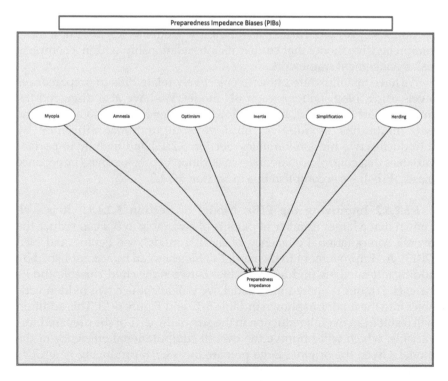

Figure 6-12. Logistic regression BN model of preparedness impedance biases (PIBs).

Table 6-9. Variables of Preparedness Impedance Biases (PIBs)
in a Logistic Regression Model.

Variable	Group	Comments
Myopia	Biases	Each random variable will describe the degree of the impedance of the bias. The states of the random variable are controlled by other factors. See Section 7.3.2.
Amnesia		
Optimism		
Inertia		
Simplification		
Herding		
Preparedness Impedance Biases (PIB)	Preparedness	This variable represents the preparedness level as controlled by the different PIBs.

in Section 3.2.4.3 that will include additional controlling variables. In Section 7.3.2, we will further use the PIBs' model as a subset of a more comprehensive model that studies its interrelationship within a complete risk management framework.

We have just illustrated how to objectively relate PIBs to preparedness levels using a logistic regression BN model. We have also discussed the limitations of the model. In the following sections, we try to address the two limitations as follows: improve computational efficiency by introducing two hidden variables, Section 3.2.4.2 and include important variables that control preparedness in addition to preparedness impedance biases (PIBs); we accomplish this in Section 3.2.4.3.

6.3.2.4.2 Improving the PIBs' Model of Section 3.2.4.1. It is well known that a larger number of parents of a variable in BN can reduce the overall computational efficiency of the BN model; see Fenton and Neil (2013). An improvement to the model's efficiency can be accomplished by adding intermediate (hidden) variables between the child variable and its parents. For our preparedness model, we will introduce two hidden variables into the model, as shown in Table 6-10 and Figure 6-13. This addition will result in an overall reduction in the size of the CPT of the preparedness variable, which will improve the overall computational efficiency of the model. Given the original large preparedness CPT, producing new CPTs for the additional hidden variables and the preparedness variable should be straightforward.

Table 6-10. Additional Variables of the Preparedness Impedance Biases (PIBs) BN Model.

Variable	Group	Comments
Hidden Variable 1 *Hidden Variable 2*	Hidden intermediate variables	The CPTs of each of these variables have only three parents. As such, the computational efficiency of the whole model should be more efficient than that of the original model of Figure 6-12.

6.3.2.4.3 Complete Preparedness BN Model. The six PIBs of the Ostrich Paradox, Section 2.7, are necessary, but not sufficient, parameters that would control the preparedness to hazards. For a more comprehensive objective model of preparedness to hazards, additional variables to the model of Figure 6-12 and its more efficient version of Figure 6-13 are needed. Table 6-11 shows some of the additional and important prepared-ness-controlling variables, whereas Figure 6-14 shows how these additional

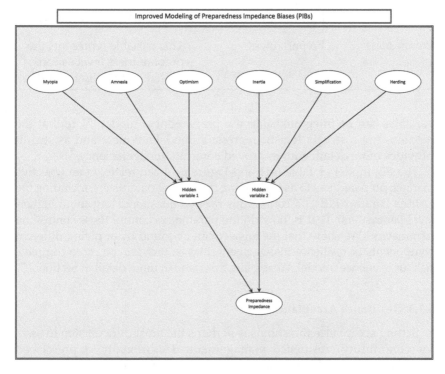

Figure 6-13. Logistic regression BN model of preparedness impedance biases (PIBs).

Table 6-11. Variables of the Comprehensive Preparedness BN Model. Additional variable Descriptions can be found in Tables 6-9 and 6-10.

Variable	Group	Comments
Infrastructure	Parameters that control preparedness, in addition to PIBs.	Adequacy of different infrastructure preparedness levels.
Budget		Available budget to meet essential costs.
Resources		Measure of available resources to be adequately prepared: human, equipment, and others.
Other		Any other variable that controls preparedness.
Hidden variable 3	Intermediate hidden variable that is introduced to improve the computational efficiency of the model.	
Preparedness	Preparedness	This variable represents the preparedness level as controlled by the different PIBs.

variables can be integrated into the preparedness model. Note that the model is not a strictly logistic regression model anymore and as such it provides more detail, with improved computational efficiency.

The BN model of Table 6-11 and Figure 6-14 can relate, in an objective manner, preparedness to its underlying controlling parameters, including the Ostrich Paradox PIBs. Of course, this raises a question: How about higher-level parameters? That is, how do the parameters control those controlling parameters? We show that the answer can be found by applying different components of resilience management, that is, providing a more complete risk (or resilience) model. We explore this issue in more detail in Section 7.3.

6.3.3 Resilience Acceptance

Setting acceptance thresholds is perhaps the most difficult step in asset (or community) resilience management. Traditionally, a prescribed acceptance threshold was adopted in civil infrastructure projects and

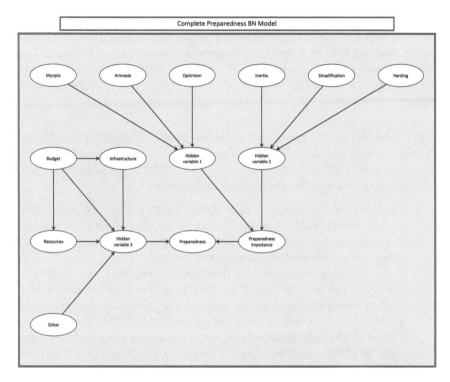

Figure 6-14. Complete BN model of preparedness impedance biases (PIBs).

based on a reasonable value. This practice aimed mostly at ensuring safety. With the advent of risk-based paradigms, where project decisions are based on safety as well as cost (both life cycle costs and initial capital expenditures), setting risk acceptance thresholds has become more difficult. The same holds true for resilience. Where do we set, objectively, a reasonable acceptance threshold for the resilience of a particular asset or particular community? The answer to this question is as important as defining and assessing resilience. For without setting a threshold, resilience improvement projects could be either unnecessarily costly or result in reduced performance (see Figure 6-42). At the time of this writing, there are few objective methods available for setting reasonable resilience acceptance thresholds, but some general suggestions can be made regarding methodologies for setting them.

Acceptance step involves relating an assessed resilience value, $Res|_a$, to a resilience acceptance threshold, $Res|_{acc}$, formally:

$$Res|_a \leq Res|_{acc} \qquad (6\text{-}17)$$

Depending on the outcome of the logical result of Equation (6-17), whether it is *True* or *False*, the decision maker would execute a particular policy, for example to retrofit the system on hand or to do nothing. The logical process of Equation (6-17) can be objective or subjective, depending on the assessment methods and the preferences of the decision maker.

6.3.3.1 Subjective Acceptance: Resilience Itself. Subjective acceptance thresholds will use descriptive expressions such as {*High*} or {*Medium*}. Equation (6-17) is then applied using one of two possible scenarios, depending on the assessment method and results:

1. If the assessed resilience is described subjectively, for example from a subjective set such as {*High, Medium, Low*}, then the logical process of Equation (6-17) is straightforward and the decision maker can either accept or reject the assessment results.
2. If the assessed resilience is described objectively, for example, $Res|_a = 25$ (on a scale from 0 to 100), then the decision maker will need to transform the objective value of $Res|_a$ (25 in our example) into a subjective value before processing Equation (6-17). In our example, a value of 25 might be translated into a subjective value of *Low* on the scale of {*High, Medium, Low*}. Processing Equation (6-17) is now possible.

Kennett et al. (2011a, b, c) used this acceptance process while addressing the state of resilience of mass transit stations, tunnels, and buildings, respectively.

6.3.3.2 Objective Acceptance. Objective acceptance still uses Equation (6-17) for the acceptance processes. In this case, both sides of the inequality are objective. There are several possibilities in applying the inequality, and these are as follows:

Recovery Time: In some situations, the decision maker desires to base resilience-related decisions on *Recovery Time*, for example, *Recovery Time* = 3 days. If an assessment process produces *Recovery Time* = 1.5 days, then the state of resilience is accepted.

Operational Level (Continued Operations): Another possible parameter for use in Equation (6-17) is a particular operational level during and after the event, *Op Level* = 9 (on a scale from 0 to 10). If an assessment process produces *Op Level* = 5.3, then the state of resilience is rejected.

Objective Resilience: From Figure 6-2, resilience can be expressed as a function of both *Op Level* and *Recovery Time* such that

$$Res = f(Recovery\ Time, Op\ Level) \tag{6-18}$$

If the decision maker wants to use *Res* as the basis for processing Equation (6-17), then both $Res|_{acc}$ and $Res|_a$ should be used to process Equation (6-17). For example, the decision maker can set $Res|_{acc} = 75$ (on a scale of 0–100). Thus, if an assessment process produces $Res|_a = 43$, then Equation (6-17) would guide the decision maker to reject the resilience state.

6.3.3.3 Resilience Acceptance: A Closer Look. The just-discussed methods for setting an acceptance threshold have one thing in common: the process of setting the thresholds themselves is rather arbitrary. This is true for both subjective and objective thresholds. Thus, in our objective examples, setting an acceptance threshold as *Recovery Time* = 3, *Op Level* = 9, or $Res|_{acc} = 75$ was arbitrary. No objective rules govern these choices. Of course, these arbitrary acceptance thresholds would be based on the desire of stakeholders to limit potential damaging consequences. If there are no other controlling factors, this approach would be a reasonable approach, albeit with some degree of subjectivity.

In practice, there is another major factor that would render the aforementioned approach unrealistic: costs. Note that in all of our acceptance methods so far, we did not account for costs. Realistically, setting a desired acceptance threshold (*Recovery Time*, *Op Level*, or $Res|_{acc}$) will always have an underlying cost. If the cost of the desired acceptance thresholds is not affordable, then these desired thresholds need to be readjusted to meet affordable cost levels. Perhaps *cost is the ultimate acceptance threshold*.

6.3.3.4 Two Lemmas of Resilience Acceptance. Discussions of Sections 3.3.1 and 3.3.3 lead us to the following two essential lemmas of resilience acceptance:

> Lemma A.1: Setting a consistent resilience metric for both assessment outcome and acceptance thresholds is needed.
> Lemma A.2: Costs need to be considered while setting acceptance thresholds.

Reflecting on the aforementioned two lemmas, we reach an important finding: by considering costs as an outcome while preforming both assessment and acceptance, we are really performing a risk rather than resilience operations, because risk, not resilience, involves itself with costs.

6.3.3.5 Case Study: Asset Acceptance Thresholds

6.3.3.5.1 Overview. As the name implies, resilience (or risk) acceptance is all about decision-making by the pertinent decision maker. It often involves choosing an optimal policy from an available set of different policies, given a set of acceptance thresholds. Because of this, this acceptance model is an influence diagram (ID) with a single decision node/variable that will comprise different available decisions (policies). The model will include three utility nodes that correspond to *Risk*. It will include a single variable that will represent *Resilience*. Given the fact that a policy that achieves optimal risk does not need to be the same policy that achieves optimal resilience, the model can have two strategies: one that results in an optimal *Risk* and another that results in optimal *Resilience*; see Section 2.5.4.1 for more discussions on policy versus strategy. We need to emphasize that all policies and strategies should always be consistent with pertinent acceptance thresholds.

6.3.3.5.2 Model. The model under consideration is a model for small business in a wildfire-prone area. The business decision makers need to address the wildfire hazard by choosing a decision (policy) out of several available decisions (policies). The available policies are the following:

1. Retrofit/harden the physical plant by introducing fire-resistant measures to reduce or eliminate vulnerability to wildfires; see Gromicko (2006) or FEMA (2020).
2. Improve resources, such as in-house fire sensors, fire drill exercises, and/or fire awareness campaigns.
3. Follow both 1 and 2 policies.
4. Provide for adequate wildfire insurance for business.
5. Move business away from the wildfire-prone area.
6. Do nothing.

In addition to the decision variable, the model is a BN-based model with three utility variables and nine random variables as explained in Table 6-12 and Figure 6-15. The prefix "A" in front of a variable in this model indicates that it is an asset-based variable. Note that the *A-Resilience* variable is a random variable with two parents: *A-Recovery time* and *A-Overall operations*. These are the two main resilience-controlling parameters. The *A-Risk* utility is a function of two utilities: *A-Retrofit costs* and *A-Loss of operations* costs. This particular separation between variables representing risk and resilience might produce different optimal policies for optimal risk and resilience. This example shows that policies for optimal risk might be different from policies that produce optimal resilience.

6.3.3.6 Case Study: Community Acceptance Thresholds

6.3.3.6.1 Overview. The model of Section 3.3.5 investigated the policies and strategies of a small business asset decision maker to enhance the asset's wildfire risk and/or resilience performance. We focus our attention now on the study of community decisions (policies) that would enhance the community's risk and/or resilience behavior in response to wildfire.

6.3.3.6.2 Model. Our first step in building a suitable model for such a study is to recognize that a community is the integral of all of its assets. So, it is natural to use the asset model of Section 3.3.5 as a base for the model of the community. In addition to the asset model (or models for multiple

Table 6-12. Description of Variables of the Asset Acceptance Model.

Variable	Group	Comments
A-Retrofit/enhance	Hidden variable	Describes levels of hardening/retrofit against wildfire
A-behavior during an event	Hidden variable	Describes the performance of an asset during a wildfire event
A-physical condition after an event	Hidden variable	Describes the performance of an asset after a wildfire event
A-Overall operations (during/after an event)	Resilience component	Describes the overall performance of an asset during and after a wildfire event
A-Operations during recovery	Hidden variable	Describes the performance of an asset during recovery from a wildfire event
A-Resources	Resilience component	Overall conditions and performance of resources
A-Recovery time	Resilience component	Duration of recovery
A-Resilience	Resilience	Estimate of asset resilience
A-Retrofit costs	Costs (risk)	Cost of retrofits/hardening
A-Loss of operations costs		Cost of loss of operations because of a wildfire event
A-Risk (total costs)		Overall costs
Wildfire	Hazard	Intensity of a wildfire event
A-Decision	Decision variable	Decision variable with six possible decisions/policies

assets) within the community, we need to add community-level variables to the model. A separate community-level decision variable is an obvious addition. The community decision variable will include the following available decisions (policies) to the community decision maker:

1. Introduce a set of laws and/or regulations that include enforcements, taxes, zoning laws, guidelines, or statutes that aim at counteracting damage from wildfires in the communities.
2. Enhance wildfire-related community resources.

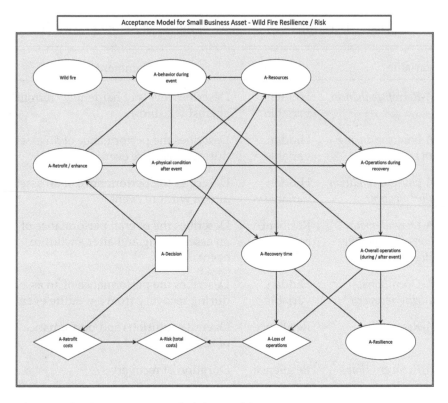

Figure 6-15. Asset acceptance decision model.

3. Embark on infrastructure repair/retrofit efforts to mitigate wildfire-related damage.
4. Enhance potential recovery efforts.
5. Do nothing (perhaps owing to a lack of budget).

The next step is to link asset-related variables to communitywide-related variables. In the following model, we consider only a single variable within the community. In more realistic settings, the model should include as many assets as needed to produce realistic and accurate results. Also, following the approach of Section 3.3.5, we separate the community resilience variable from the community risk variable in recognition of the possibility of having two different strategies that would produce optimal risk and optimal resilience, respectively.

The proposed model is shown in Table 6-13 and Figure 6-16. Again, a prefix "A" in the variable's name refers to an asset-based variable, whereas a prefix 'C' in the variable's name refers to a community-related variable. Note that a new utility variable is added, *C-Risk*, which represents the overall community risk. Similarly, the *C-Resilience* variable presents the overall community resilience.

Table 6-13. Description of Variables of the Asset Acceptance Model.

Variable	Group	Comments
A-Retrofit/enhance	Hidden variable	Describes levels of hardening/ retrofit against wildfire
A-behavior during an event	Hidden variable	Describes the performance of an asset during a wildfire event
A-physical condition after an event	Hidden variable	Describes the performance of an asset after a wildfire event
A-Overall operations (during/after an event)	Resilience component	Describes the overall performance of an asset during and after a wildfire event
A-Operations during recovery	Hidden variable	Describes the performance of an asset during recovery from a wildfire event
A-Resources	Resilience component	Overall conditions and performance of resources
A-Recovery time	Resilience component	Duration of recovery.
A-Resilience	Resilience	Estimate of asset resilience
A-Retrofit costs	Costs (risk)	Cost of retrofits/hardening
A-Loss of operations costs		Cost of loss of operations because of a wildfire event
A-Risk (total costs)		Overall costs
Wildfire	Hazard	Intensity of a wildfire event
A-Decision	Decision	Decision of the asset decision makers. It contains six possible decisions/policies
C-Decision	Decision	Decision of the community decision makers. It contains five possible decisions/policies.

(Continued)

Table 6-13. (*Continued*) Description of Variables of the Asset Acceptance
Model.

Variable	Group	Comments
C-Laws/regulations	Policy component	States of this variable represent the complexities of the different laws and regulations produced by the community.
C-Resources	Policy component	States of this variable represent the level of efforts undertaken by the community to enhance its resources to counteract wildfires.
C-Repairs	Policy component	States of this variable represent the level of efforts undertaken by the community to enhance its infra-structure (roads, bridges, public buildings, fire equipment, etc.) to counteract wildfires.
C-Recovery	Policy component	States of this variable represent the level of efforts undertaken by the community to enhance its recovery operations (first responders, reconstruction efforts, etc.) to counteract wildfires.
C-Risk (total costs)	Costs (risk)	Overall community risk (costs)
C-Resilience	Resilience	Overall community resilience

6.3.4 Resilience Improvement/Treatment

The resilience improvement/treatment component of resilience management becomes necessary when the resilience of a particular asset or community is determined to be below the pertinent resilience acceptance threshold. The resilience improvement component includes three phases: baseline improvements, project plans and prioritizations, and project execution. Similar phases are usually followed for improvements/treatments that are based on other infrastructure paradigms such as risk, reliability, or sustainability. For now, we concentrate on resilience improvements.

Resilience improvements/treatments include three phases as illustrated in Figure 6-17. Phase I resilience improvements attempt the

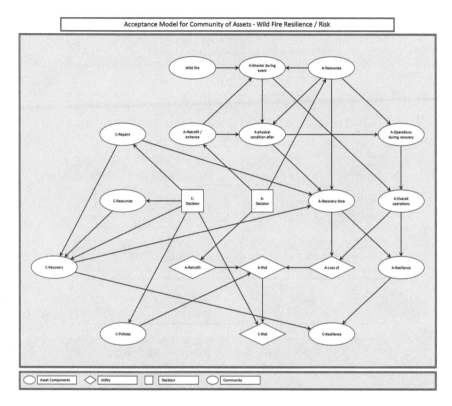

Figure 6-16. Community acceptance decision model.

most efficient type of improvement. There are four possibilities as explained in Table 6-14. If the outcome of Phase I is to embark on a resilience improvement effort, decision makers then move into Phase II, which is preparing different potential resilience improvement projects that need to account for the potential for the improvement of some or all of the 4Rs. Many times, decisions are made to physically improve facilities or communities. Studies are needed to determine a balance between the improvements in the 4Rs to achieve the optimal benefit-to-cost ratio. After several potential candidate improvement projects are assembled, a project prioritization effort is needed to choose the most suitable and cost-effective project to achieve the required resilience level. Project prioritization is not an easy task. It must consider cost, benefit, project scale, and life cycle. See Fenton and Neil (2013) for a theoretical background and Ettouney and Alampalli (2017a, b) for practical examples. After deciding on the most suitable project for resilience improvements, Phase III entails project execution. Figure 6-17 shows the three phases of resilience improvement.

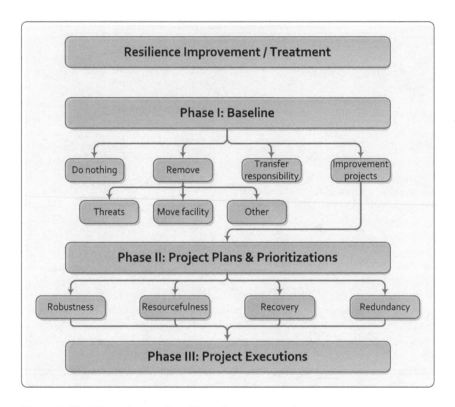

Figure 6-17. Three phases of resilience improvement/treatment.
Source: Ettouney (2014).

6.3.4.1 Many Dimensions of Treatment/Mitigation. Because there are several dimensions of resilience, as discussed in Section 1.4, it is reasonable to expect that resilience treatment/improvements will also possess several dimensions. We note that multidisciplinary, multihazard, and cascading considerations and preparedness dimensions can have important effects, perhaps controlling one, on how resilience is treated and/or improved. Obviously, each of these dimensions is complex in its own right. Because of this, we discuss each of them separately, as they apply to resilience, later in Sections 4, 5, 6, and 6.4.

6.3.5 Resilience Monitoring

6.3.5.1 Overview. As in every aspect of life, the resilience of assets and communities degrades with the passage of time. This degradation affects all the 4Rs; see Section 3.1. For example, the physical deterioration of facilities is well documented (see Agrawal and Kawaguchi 2009). Operations can also degrade with the passage of time, perhaps because of

Table 6-14. Decision Options for Phase I of Resilience Improvements/
Treatments.

Type of action	Comments
Do nothing	Some of the reasons can be as follows: • The margin between accepted threshold and actual resilience ratings is minimal. • Budget constraints. • Other subjective circumstances such as a lack of confidence in the analysis, similar improvement is planned in the near future, or a subjective feeling that the postulated threat/hazard is too rare to be considered.
Remove potentially damaging events	If threats or hazards can be removed entirely, do so. Man-made hazards are prime examples. Natural hazards may be difficult to anticipate, but the entire facility can be moved away from exposure to a potential natural hazard.
Transfer responsibility	Reasonability for potential interruptions of operations might be transferred by opting for insurance. In the case of many civil infrastructure, such as bridges or tunnels, this option is not feasible.
Improve/treat deficiencies	Improvements include physical retrofits and/or operational improvements.

increased demands, personnel retirement, or budget constraints. Because of this, and to avoid having a state of resilience below an acceptable threshold, see Figure 6-43, resilience needs to be monitored as a component of the resilience management process. In addition, ReMo techniques need to cover all the 4Rs of resilience; see Betti (2022) and Flanigan et al. (2022). Some of the techniques are shown in Table 6-6.

6.3.5.2 Case Study: Resilience Monitoring of Bridge Subjected to Scour

6.3.5.2.1 Overview. Scour hazard is one of the most damaging hazards that can impact a bridge; see Annandale (2006) or Ettouney and Alampalli (2011a). One of the attributes of scour is that it is not easy to detect it visually, and in many situations, it requires special monitoring techniques; see Ettouney and Alampalli (2011a). Table 6-15 shows some ReMo techniques categorized by resilience 4R components. In this example, we look at the *ReMo* of a bridge that is subjected to scour hazard.

Table 6-15. Examples of ReMo Techniques.

Component	Monitoring techniques
Robustness	Structural health monitoring methods (see Ettouney and Alampalli 2011a, b).
Resourcefulness	Training, courses, management, and operational auditing for compliance.
Recovery	Auditing, tabletop, and other types of exercises and drills. Pertinent inspection and maintenance processes.
Redundancy	For physical redundancy, use structural health monitoring methods. See Ettouney and Alampalli (2011a, b). For operational redundancies, use techniques of monitoring resourcefulness and recovery.

When planning for a *ReMo* effort, the decision maker will need to accommodate the three lemmas of Flanigan et al. (2022) as well as the monitoring approach of Betti (2022). The monitoring effort will need to accommodate the following:

- Principle of Resilience Monitoring, *PRM*, as introduced by Flanigan et al. (2022), governs the basis of ReMo. The *PRM* states that both performance quality, *Q*, and estimated recovery time, *T*, need to be monitored. The *PRM* also indicates that monitoring a single bridge (or asset) is necessary but not sufficient for a proper *ReMo*; the rest of the bridge network (or the whole community), or parts of it, needs to be monitored too, as shown by Betti (2022).
- Both foundation and superstructure monitoring are needed. This is Lemma 1 from Flanigan et al. (2022).
- There is a need for other monitoring types, in addition to visual inspection, *VI*. For example, structural health monitoring, *SHM*, may be needed. This is Lemma #2 from Flanigan et al. (2022).
- Visual inspection is always needed. This is Lemma #3 from Flanigan et al. (2022).

In the following model, we show how to accommodate all of the aforementioned essential principles and lemmas in a CG model to monitor bridge resilience.

6.3.5.2.2 Model. The variables of the model for monitoring bridge resilience regarding scour are shown in Table 6-16, and the model itself is

Table 6-16. Description of Variables of the Resilience of Bridge against Scour.

Variable	Group	Comments
Resource Monitoring	Monitoring activity	Adequacy of resources such as emergency rehabilitation efforts or underwater operations. In addition, it includes monitoring of all other parameters that might control resources, such as available budgets, human resources, equipment, and others.
SHM	Monitoring activity	Use SHM technologies for monitoring such as soil conditions under foundations, structural stability, and strains in both foundations and structures.
Scour	Controlling hazard	
VI	Monitoring activity	Regular and special visual inspection procedures
Foundation	Infrastructure	Conditions of bridge foundations
Superstructure	Infrastructure	Conditions of bridge superstructures
Resource deployment	Resilience component	Efficiency of resource deployment in case of emergency
Bridge network	Infrastructure	Includes roadways, other bridges, tunnels, and so on. Note that *Bridge Network* is linked to *Performance* via a nondirectional link. This renders this model as a chain graph (CG) model.
Performance	Resilience component	Including adequacy of detours, if applicable/needed)
Network monitoring	Monitoring activity	Adequacy of bridge networks
Recovery	Resilience component	Efficiency of recovery operations
Resilience	Resilience component	Overall bridge resilience in case of scour events
Improvements	Infrastructure	Improvements in bridge foundations and/or superstructures owing to observations of *VI* and/or *SHM*
Resources	Infrastructure	Adequacy of needed resources

displayed in Figure 6-18. The model is a CG-based network. The links in the model are mostly directional. However, there is a nondirectional link in the model between *Bridge Network* and *Performance*. The model can produce an estimate of bridge resilience to a scour event because of the observations (evidence) of several monitoring variables: *VI, SHM, Resource Monitoring*, and *Network Monitoring*. As stated previously, *VI* and *SHM* are necessary but not sufficient for adequate bridge *ReMo*. *Resource Monitoring* and *Network Monitoring*, as developed by Betti (2022), are also needed for adequate bridge *ReMo*.

The *ReMo* model of Table 6-16 and Figure 6-18 gives a snapshot estimate of bridge scour resilience. As noted in Section 3.5.1, the severity of hazards changes over time. Also, infrastructure properties degrade over time. Because of this, *ReMo* will need to be performed either continuously or intermittently. Also, forecasting *Resilience* is also needed for planning purposes. This means that the model of Table 6-16 and Figure 6-18 needs to be time-dependent. We discuss this subject next.

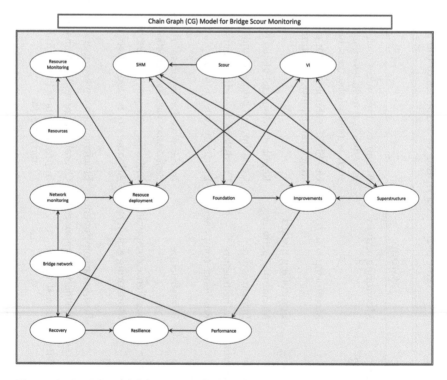

Figure 6-18. CG Model for ReMo of a bridge that is subjected to scour.

6.3.5.3 Case Study: Resilience Monitoring of Bridge Subjected to scour as a function of Time

6.3.5.3.1 Overview. This model will expand on the model of Section 3.5.2 to show how *ReMo* would observe temporal changes of bridge *resilience* as well as its components. We utilize the principles of dynamic chain graphs (DCG) to achieve this goal. The model of Section 3.5.2 will need to change to accommodate the DCG computational needs as follows:

- Set of time link nodes (TLNs) will be introduced in the model. TLNs will represent the temporal changes in some infrastructure properties.
- We will assume that all monitoring activities are done intermittently. Because of this, we introduce two time slices describing the model as follows:
  - Time slice to model the times at which monitoring activities occur.
  - Time slice to model the instances when no monitoring activities occur.

The main goal in this model is to show how *Resilience* degrades as time passes, which generates a need for adequate monitoring of the four main variables that control the bridge resilience regarding scour hazard: *Foundations, Superstructure, Resources,* and *Bridge Network.*

6.3.5.3.2 Model. The variables of Table 6-16 are used for the time slice when monitoring activities occur, as shown in Figure 6-19. The only modification is in the description of some nodal groups as in Table 6-17. For the time instance when no monitoring activities occur, some of the variables of Table 6-16 are not used, as shown in Figure 6-20. We offer a summary of the process next.

6.3.5.3.3 Summary of the Resilience Forecasting Process Using CGN and ReMo. Let us consider a situation where monitoring is performed every $N = 5$ years and resilience forecast is needed on a yearly basis.

The CGN model of Table 6-16, Table 6-17, Figure 6-19, and Figure 6-20 can be utilized as follows:

1. Develop a *ReMo* CGN model using the template model of Table 6-17 and Figure 6-19, and adjust the template as needed. This is a time slice, T_0.
2. Use appropriate observations of the monitoring variables of Tables 6-16 and 6-17 and find corresponding probabilities, including estimates for *Resilience* at time T_0, Res_0.
3. Use adequate transformation matrices, see Section 2.5.5, for the computed probabilities of the variables at different *TLNs* to develop soft evidence values at the next time step (next year in this example), T_1.

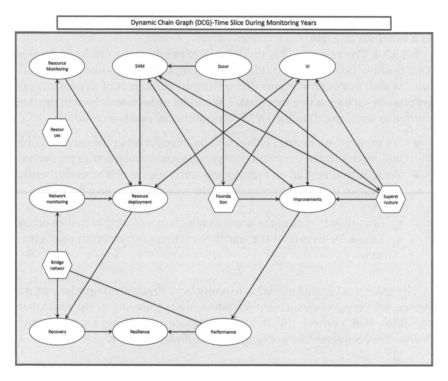

Figure 6-19. Time slice of the DCG model of ReMo of bridge-monitoring years.

4. Use the intermediate time slice model of Figure 6-20 and the soft evidence developed in Step 2 to evaluate forecasted probabilities at all nodes. We now have a forecast for *Resilience* at time T_1 (first year), Res_1.
5. Repeat Steps 3 and 4 to obtain forecasted probabilities at all nodes, including *Resilience*, for each year up to year 5.

Table 6-17. Description of Modified Variable Groups of the DCG Model of Resilience of Bridge against Scour.

Variable	Group	Comments
Foundation	Time link node	These four nodes are infrastructure nodes, as in Table 6-16. However, they are also time link nodes (TLNs) where their probabilities will be computed in time future slices using transfer matrices. Note also that these TLNs are present across all time slices.
Superstructure	Time link node	
Bridge network	Time link node	
Resources	Time link node	

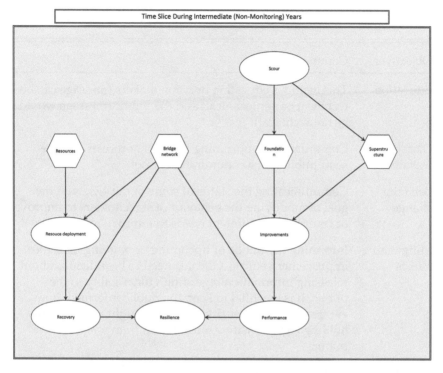

Figure 6-20. Time slice of the DCG model of ReMo of bridge-non-monitoring years.

6. For year 5, use the model in Figure 6-19 to allow for monitoring observations to be used in the model. Recall that the actual monitoring activity is performed at 5-year intervals. Solve the model given these observations to find the probabilities of *Resilience* at year 5, Res_5.

7. Repeat Steps 1 through 6 to get the required forecasts for the following 5-year intervals.

6.3.6 Resilience Communication

The final, and perhaps most important, component in resilience management is resilience communication, which is the bridge between the technical/professional community, management and decision makers, the legislature, funding sources, and the public. There are several objectives of resilience communication, some of them are presented in Table 6-18.

Similar to risk communication, see Lasswell (1948), and Gutteling and Wiegman (1996), resilience communication can be subdivided into five essential modules: originator (source), content (message), recipient (target audience), medium (channel), and objectives (destination). The five

Table 6-18. Objectives of Resilience Communication.

Objectives	Comments
Education	This includes educating decision makers (message can be technical or nontechnical) and the public (message would be nontechnical).
Disaster warnings	Communicating upcoming hazards or threats and the susceptibility of a community or asset.
Behavior change	Communicating the status of some of the 4Rs, with the goal of modifying the behavior of stakeholders to improve or comply with different needs to improve resilience.
Mitigation efforts	Informing the public of upcoming or ongoing resilience improvement efforts. Officials need to be judicious about releasing information regarding vulnerability to the public. It is essential to keep the public informed, however; sometimes security situations might require withholding some sensitive information from release to the public.
Funding	Relay funding needs.

modules should be interrelated as shown in Figure 6-21. As such, the resilience communication modules should form a complete graph. A successful resilience communication plan will need to accommodate the five modules in a careful and well-thought-out manner. Obviously, resilience communication plans will/should differ greatly depending on the nature of any of the aforementioned five modules, as well as the nature of the type and size of the asset/community and its stakeholders.

6.3.6.1 Avoiding Ostrich Paradox Preparedness Impedance Biases
6.3.6.1.1 Overview.   Our discussion of the Ostrich Paradox in Section 2.7 explained the different PIBs. PIBs are related directly to resilience, or risk, communication, because effective resilience communication can help in reducing the effects of biases and increase public reception to preparedness efforts and costs. In this section, we offer an example showing how modules of communication (see the Laswell model of Figure 6-21) can objectively measure the effectiveness of a communication message in reducing two of the six PIBs. We use the public indifference game process of Section 2.6 to solve this problem.

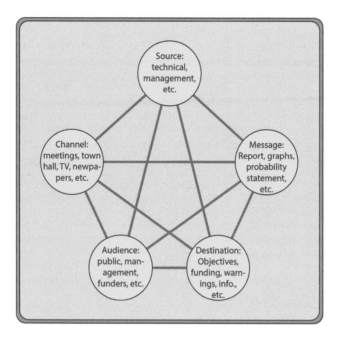

Figure 6-21. Resilience communication modules, based on the Laswell Model. Source: Ettouney (2014).

6.3.6.1.2 Dominant Equilibrium Strategy. In our example, we assume that a public organization (a source of communication, as per the Laswell model) in a flood-prone region wants to develop a communication message to the public (receivers, as per Laswell model). The organization realizes that to improve the state of preparedness in the region, it needs to reduce two of the PIBs: public myopia and public amnesia. The organization needs to choose an appropriate mix of messaging strategies: a messaging strategy that emphasizes the dangers of future floods or a messaging strategy that emphasizes historical floods. A public survey revealed a binary public sentiment (accept message/reject message) for each of the two messaging strategies. In the same survey, the frequency of each message as related to public sentiment was also observed. The resulting payoff (on a scale of 0 to 100) is shown in Table 6-19.

According to the public indifference game, Section 2.6.4, the optimal strategy mix of the public organization occurs when a long-term equilibrium of dominant strategies for both the organization and the public occurs at the same outcome. Studying Table 6-19, we find that the dominant strategy of the public organization is the message to avoid amnesia (80), whereas the dominant strategy for the public sentiment is to

Table 6-19. Payoff Table for Ostrich Bias Avoidance: Public indifference to a message with a Dominant Strategy Equilibrium.

| | | Public Sentiment | | | |
| | | Accept | | Reject | |
Payoff for the strategies of the two players		Message frequency	Degree of acceptance	Message frequency	Degree of rejection
Preparedness message	Avoid myopia: point to future forecasts	75	25	60	10
	Avoid amnesia: point to past experiences	80	80	65	45

accept (80). Both dominant strategies coincide in the same cell in the payoff table, which produces a dominant strategy equilibrium. This means that the long-term equilibrium for the organization is to emphasize avoiding amnesia and that the public will accept such a message, because it is also the dominant strategy for the public. There is no need for the organization to mix its messaging strategies: just concentrate on an avoid amnesia message and the public will accept such a message in the long run.

6.3.6.1.3 Nash's Equilibrium Mixed Strategy. Now let us consider a more complex case. Assume that the payoff table is as shown in Table 6-20. The new payoff results do not produce dominant equilibrium strategies for both players (public organization and public sentiment). Thus, the public organization will need to use a mix of strategies, that is, mixing the messages. To produce a Nash equilibrium mixed strategy, which will be the optimal messaging mix, such a mix will need to produce a public indifference to the messaging strategies of the public organization; see Section 2.6.4. Employing Equation (6-16), relationships between the frequency of messaging to avoid myopia, public acceptance payoff, and public rejection payoff are shown in Table 6-21 and Figure 6-22. Public indifference will occur when the two expected utilities of the two public sentiments are equal; see Equation (6-16). This will happen when the frequency of messaging to avoid myopia is 40% and messaging to avoid amnesia is 60%.

Table 6-20. Payoff Table for Ostrich Bias Avoidance: Public indifference
to a message with Nash's Equilibrium and Mixed Strategy.

		Public sentiment			
		Accept		Reject	
Payoff for the strategies of the two players		Message frequency	Degree of acceptance	Message frequency	Degree of rejection
Preparedness message	Avoid myopia: point to future forecasts	50	25	65	55
	Avoid amnesia: point to past experiences	80	50	60	30

Table 6-21. Public Reactions to Myopia vs Amnesia Avoidance.

Frequency of messages that point to future forecasts (avoid myopia)	Public accept	Public reject
0%	50.00	30.00
10%	47.50	32.50
20%	45.00	35.00
30%	42.50	37.50
40%	40.00	40.00
50%	37.50	42.50
60%	35.00	45.00
70%	32.50	47.50
80%	30.00	50.00
90%	27.50	52.50
100%	25.00	55.00

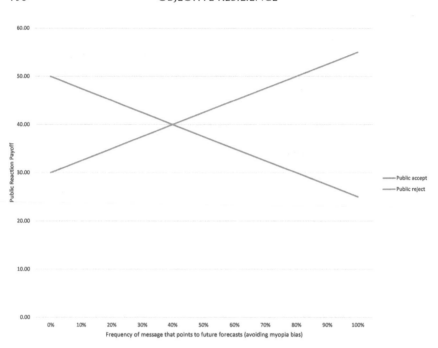

Figure 6-22. Public reactions to myopia versus amnesia avoidance: mixed strategy and Nash's equilibrium.

This is the optimal strategy mix (Nash's equilibrium) of the two messages that the public organization needs to use to produce the optimal messaging result.

6.4 RESILIENCE AND ITS MULTIDISCIPLINARY UNDERPINNINGS

6.4.1 Overview

Achieving resilient systems requires the cooperation and interaction of many disciplines and/or agencies. We considered multidisciplinary (interagency) efforts as one of the dimensions of resilience in Section 6.1.4. Of course, the type and number of disciplines/agencies that need to be considered while addressing the resilience of a particular system is context-dependent. For example, addressing the resilience of a small community at risk of flooding will require different disciplines/agencies from those that are needed by a large business that is exposed to wildfires.

We address two objective multidisciplinary resilience considerations next. In Section 6.4.2, we offer some insight into modeling and assessing the efficiency of a multidisciplinary setting. In Section 6.4.3, we offer an example of an objective evaluation of the multidisciplinary efforts of a typical emergency management setup.

6.4.2 Objective Multidisciplinary Effectiveness

Our main concern here is to try to express the multidisciplinary nature of any system in a form that is amenable to objective processing. Following Mohammadfam et al. (2014), GN modeling can be used to achieve this goal. Using a GN to model the links among different disciplines in a system makes it easy to objectively assess the efficiency, strengths, and weaknesses of the multidisciplinary system in an objective manner. This will make it easy to improve the multidisciplinary effectiveness and, thus, improve the overall resilience of the system. We will explore this subject later in Sections 6.4.3 and 6.7.5.

Recall that GN topology consists of nodes (vertices) and links (edges). In modeling multidisciplinary systems as GN, we use the following guidelines:

1. Discretize the system into a set of nodes. Each node represents a discipline, organization, office, activity, and others. The discretization can be low, medium, or high resolution, depending on the particular situation on hand.
2. Establish a set of links among appropriate nodes. The links represent potential direct communication among nodes. Within the network, there should be no link between pairs of nodes that do not have direct communication between them.
3. The properties of a link between a pair of nodes include the following: The nature of communication among the nodes. It can be a two-way communication (nondirectional link). The link can also be one-directional, from one node to the other. The importance of the link between a pair of nodes can be simulated by assigning a weight to the link. If all communication among the nodes are equally important, then a unit weight is assumed for all the links.

The properties of the resulting GN can then be used to objectively study the multidisciplinary properties of the system.

Figure 6-23 shows a simple GN model for a multidisciplinary system. The system contains five interlinked disciplines. Note that the GN is a complete graph, that is, all the nodes are connected to all other nodes. A quick analysis of this system indicates that it is well connected because all nodes are linked to one another. However, note that the degrees of all the nodes are maximum (4), which can result in some inefficiencies. Clearly, a balance between the two GN attributes (connectivity versus efficiency) is definitely needed to obtain an efficient and well-connected multidisciplinary system.

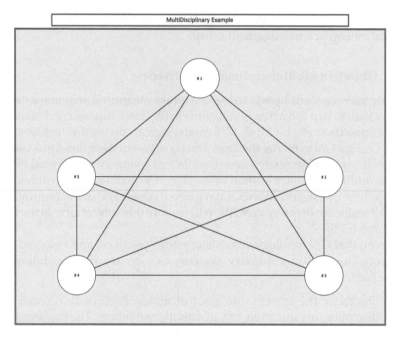

Figure 6-23. Simple example of a multidisciplinary system as a complete graph model.

6.4.3 Emergency Management is a Multidisciplinary Effort

The field of emergency management is a good example of the importance of multidisciplinary considerations for achieving resilient systems. An illustration of the essential nature of multidisciplinary underpinnings for emergency management can be found in DHSEM (2019). The phases of the State of Colorado emergency management system are as follows:

- Preparedness,
- Response,
- State mitigation section,
- State recovery officer,
- Field services section, and
- Private sector liaison.

The organization chart of the emergency management is displayed in Figure 6-24. The multidisciplinary nature of the system is evident. It is then not surprising that most, if not all, emergency management operations are multidisciplinary because the very nature of emergency management requires input from several disciplines and teams.

Following the lead of Section 6.4.2, we can model emergency management multidisciplinary systems that are similar to the ones of

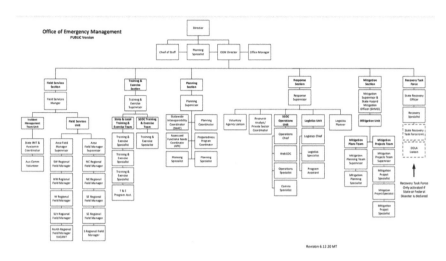

Figure 6-24. Organization chart of the Office of Emergency Management, State of Colorado.
Source: DHSEM (2019).

DHSEM (2019) and Figure 6-24 in a GN topology. For example, the organizational chart of Figure 6-24 can be simplified into a GN-based model, as shown in Table 6-22, that summarizes different nodal names, and in Figure 6-25 that shows the topology of the model. As expected, the GN model is mostly a tree structure, because it is based on an organizational chart. Utilizing similar GN models for emergency management, the stakeholders can proceed to investigate the topology of the system in an objective manner, which can lead to discovering system weaknesses and then addressing such weaknesses, which will result in an even more efficient and robust multidisciplinary system. GN properties such as the ones given in Section 6.2.4 can aid in such discovery.

6.5 THE MULTIFACETED MULTIHAZARD CONSIDERATIONS

6.5.1 Overview

Let us define a hazard as any event that might degrade the performance of an asset or a community. Considering this reasonable definition of a hazard, we note that abnormal man-made events (such as bomb blasts) or natural events (such as earthquakes or superstorms) can, of course, be considered hazards. However, we must also consider less dramatic, but equally destructive, processes, such as corrosion or normal wear and tear, as hazards. It follows that assets or communities are rarely subjected to just one type of hazard. Ettouney et al. (2005) introduced a theory of

Table 6-22. Model Variable IDs.

ID	Function/discipline
01	Director
02	Chief of staff
03	Planning specialist
04	OEM director
05	Office manager
06	Field service section
07	Training and exercise section
08	Planning section
09	Response section
10	Mitigation section
11	Recovery task force
12	Incident management team unit
13	Field services unit
14	State and local training exercise team
15	SEOC training and exercise team
16	SEOC operations unit
17	Logistics unit
18	Mitigation unit
19	Mitigation plans team
20	Mitigation project team

multihazards, postulating that different hazards in the civil infrastructure community will interact through the physical system in a parallel manner. This means that a multihazard resilience strategy needs to be implemented, as opposed to dealing with one hazard at a time. Failing to follow a multihazard resilience strategy might result in costly unintended consequences for assets and/or communities.

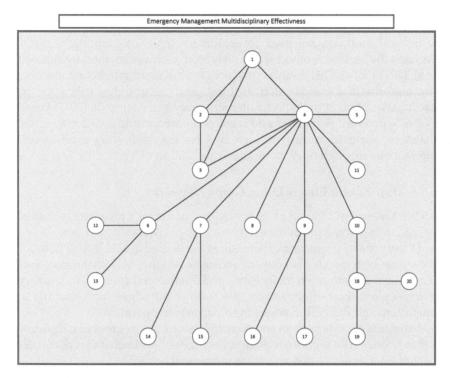

Figure 6-25. Model with mostly a tree structure.

Generalizing the multihazard efforts of Ettouney et al. (2005), Ettouney and Alampalli (2017a) introduced two theories of multihazard considerations in civil infrastructure, the multihazard physical theory (MPT) and the multihazard decision theory (MDT). Hughes (2022) introduced a third aspect of multihazard considerations: the multihazard operations theory (MOT). The MPT argues that hazards interact through the physical system under consideration. The MDT argues that a decision regarding one hazard will have an effect not only on the system's behavior to that hazard, but also on the system's behavior because of any other hazard that might affect the system. Finally, the MOT, as introduced by Hughes (2022), argues that operational efforts regarding one hazard might have an effect on operations that might affect other hazards.

Reflecting on the preceding points, we can state the following resilience multihazard lemma:

Lemma MH.1: To achieve an efficient resilient system, the decision maker needs to consider not only all pertinent hazards but also their different interactions.

6.5.2 Objective Evaluations of Multihazard Effects

Several methods are used to evaluate interactions among hazards. Perhaps the earliest method was a weighted averages method by Kennett et al. (2011a, b, c). This method produces multihazard interaction matrices for man-made hazards such as blast and natural hazards such as earthquakes, floods, and wind. Later, Ettouney and Alampalli (2017a) used PGNs to produce risk-based and resilience-based multihazard interaction matrices. We explore next the use of GNs for evaluating multihazard interactions through operations as an application of MOT.

6.5.3 Multihazard Effects Using Graph Networks

6.5.3.1 Overview. Given a GN operations model for a given organization, we can define some multihazard interaction properties as follows:

Density of MH interactions: It measures the degree of MH interactions for the whole system. A higher density indicates a higher MH interaction, thus requiring more attention to decisions and operational specifics to improve efficiency and cost-effectiveness (the ratio of all edges from hazards to operations/all POSSIBLE edges from hazards to operations).

Individual hazard effects on operations: These are the degrees of individual hazard nodes. The higher the degree, the higher the effect of that particular hazard on a system's operational management.

Multihazard Operational Matrices, $[MOM]$ and $[\overline{MOM}]$: We define the multihazard operations matrix, $[MOM]$, as a matrix with an order of N_H representing the number of hazards under consideration. The individual cells, MOM_{ij}, are evaluated as follows:

$$MOM_{ii} = N_{H_{ii}} \qquad (6\text{-}19)$$

and

$$MOM_{ij} = N_{H_{ij}} \qquad (6\text{-}20)$$

$$N_{H_{ii}} = \text{Total number of operations that relate to the } i\text{th hazard} \qquad (6\text{-}21)$$

and

$N_{H_{ij}} = $ Total number of operations that affect both the ith and the jth hazards

$$(6\text{-}22)$$

Note that, by definition, $MOM_{ij} \leq MOM_{ii}$. Also, $i \in N_H$ and $j \in N_H$ represent the rows and columns in MOM. The individual cell in MOM

indicates the relative effects of the ith hazard and the jth hazard on the overall operations in the system. Note than an $MOM_{ij} = 0$ would indicate no MH interaction in operations between the ith hazard and the jth hazard. We also note that $[MOM]$ is a symmetric matrix.

If we normalize $[MOM]$, such that the normalized matrix, $\overline{[MOM]}$, the individual normalized cells can be evaluated as follows:

$$\overline{MOM}_{ii} = 1.0 \tag{6-23}$$

and

$$\overline{MOM}_{ij} = \frac{N_{H_{ij}}}{N_{H_{ii}} + N_{H_{ij}}} \tag{6-24}$$

The resulting normalized matrix will be nonsymmetric with unit diagonals. Each of the normalized and non-normalized matrices would indicate the interactions in operations between the ith hazard and the jth hazard. A knowledge of interaction values then accounting for them in planning and executing operational processes accordingly can improve the efficiency of operations of the system.

6.5.3.2 Example: Operational Multihazard Considerations in an Rapid Transit Systems.

As an example, we develop an operational GN for an RTS that is based on the operational RTS description given by Hughes (2022). The GN nodes and their descriptions are shown in Table 6-23. The GN topology is shown in Figure 6-26. For simplicity, we consider only two hazards in the model: snow and flood. The links in the model are all directional. The nodes are combinations of hazards, operations, and generic operations that lead to specific operations.

We can develop the second-order (because we have only two hazards in this example) $[MOM]$ and $\overline{[MOM]}$ by simple counting. Note that Operations 8, 10, 16, 17, 18, 19, 20, and 21 are affected by both hazards. Even though Operation 10 is not directly affected by flood, it lies on Path $2 \rightarrow 5 \rightarrow 10$. Operations 5, 9, 11, and 12 are affected only by flood, whereas Operations 6 and 7 are affected only by snow. Thus, we find that $N_{H_{11}} = 10$, $N_{H_{12}} = 8$, $N_{H_{21}} = 8$, and $N_{H_{22}} = 12$. The non-normalized and the normalized multihazard operational matrices can be expressed as follows:

$$[MOM] = \begin{bmatrix} 10 & 8 \\ 8 & 12 \end{bmatrix} \tag{6-25}$$

Table 6-23. Key to Node Numbers and Names.

Node #	Node name	Group
01	Flood	Hazards under consideration
02	Snow	
03	Infrastructure	Generic operations
04	Maintenance	
05	Pump rooms	Operations involving mainte-nance, repair, and rehabilitation (MRR) These operations are affected directly by one or more hazards.
06	Drainage	
07	Tunnel doors	
08	Bridges	
09	Station	
10	Track	
11	Signal	
12	Power	
13	Lighting	Other operations, that is, operations that are not directly affected by hazards.
14	Vehicles	
15	Inspection	Generic operations
16	Periodic (inspectors)	Indirect MRR operations are affected directly by one or more hazards.
17	Monitoring (sensors)	
18	Standard operating procedures (SOPs)	
19	Training	
20	Interagency cooperation	
21	Outside contractors	

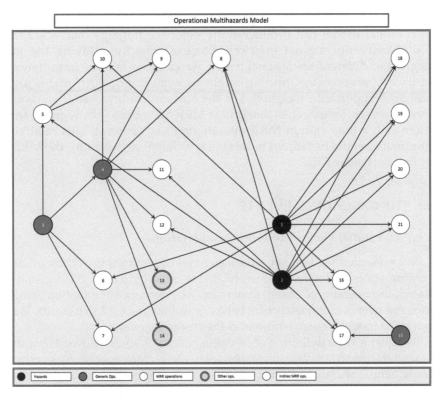

Figure 6-26. Directed graph of MH operations for an RTS.

Also, using Equation (6-24),

$$\left[\overline{MOM}\right] = \begin{bmatrix} 1.0 & 0.8 \\ 0.667 & 1.0 \end{bmatrix} \qquad (6\text{-}26)$$

respectively.

Note that the relative number of operations of flood owing to snow is slightly higher than the reverse (0.8 versus 0.667). This means that planning for snow operations would produce better payoff than planning for flood operations, assuming that the choices available are binary choices. A more illuminating process would be to assign importance weights to different nodes and links and include those weights while developing the matrices. Also, a higher and more detailed GN model is advisable because the proposed model might not be detailed enough for accurate operational planning. Finally, using more hazards in the model can result in even more efficient operational planning. However, these changes are beyond the scope of this chapter.

Before we end this section, it is worth mentioning that the GN operational model can produce more objective findings than just the multihazard operational matrices. For example, by studying the in degrees of different operational nodes, we can ascertain the importance of each operation (this finding might be even improved if nodes are assigned importance weights). For the current uniform nodal weights model, we see, for example, that bridge MRR operations (3 in degrees) are more active than station MRR operations (2 in degrees). This relative importance could be helpful in assigning different resources for the MRR of that particular RTS.

6.6　THE CASCADING EVENTS

6.6.1　In-Parallel versus In-Series Considerations

We presented the issue of multihazard considerations in Section 6.5 as considerations of different hazard-related issues (physical-based, decision-based, and operational-based) in parallel. We now turn our attention to the in-series events of infrastructure behavior as they respond to hazards. We lump the in-series considerations as the cascading events.

We offer a loose definition of cascading events (CE) as follows: *When an undesirable event affects a particular asset (or a component of an asset or community), and degrades its performance, the influences of this initial degradation propagate downstream through the asset (or community), and during such propagation, the performance of other components degrades.* The degradation process will continue serially throughout the asset (or community). The net result, when the degradation propagation finally stops, can be a severe deterioration of the operation of the asset or the community.

Following the aforementioned definition, and similar to multihazard considerations, cascading events have three different cascading modes: cascading hazards, cascading effects, and cascading failures. Unfortunately, these three terms are often used interchangeably. However, there are important differences among these three modes of cascading events that we need to be aware of, because these differences would result in different mitigation strategies for each of them. Next, we discuss each mode in depth and point out its general mitigation strategy.

6.6.2　The Three Modes of Cascading Events

6.6.2.1　Cascading Hazards.　A cascading hazard event occurs when a particular hazard would affect the whole system in such a manner that another totally different hazard occurs. The second hazard will damage the system further. The threat of additional hazards occurring might continue, depending on the overall system attributes. Two well-known cascading

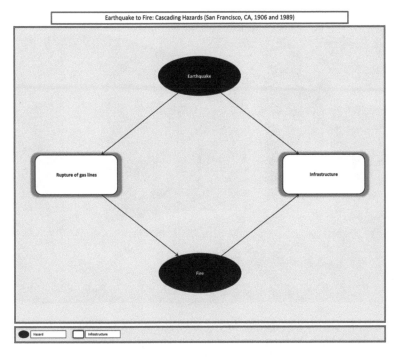

Figure 6-27. Cascading hazards, San Francisco earthquakes, 1906 and 1989.

hazard events impacted the San Francisco area in 1906 and 1989. In each of these two events, a severe earthquake hit the city. The earthquakes caused gas lines to break, which, in turn, caused a fire with catastrophic consequences; see Figure 6-27. When hurricane Sandy was battering lower Manhattan, New York, in 2012, a combination of heavy rain and strong wind caused window glass breakage. This, in turn, caused major indoor flooding of businesses and residences; see Figures 6-28 and 6-29. Also, during the same hurricane and in the same area, the resulting storm surge, coupled with inadequate protections, resulted in the flooding of subway tunnels and stations. This, in turn, caused another hazard, albeit not a natural one: the widespread interruption of subway services in the New York City area; see Figures 6-30 and 6-31.

Mitigating the effects of cascading hazards is a fairly straightforward exercise: we need to just provide adequate protection from expected hazards. Of course, protection against the topmost hazard in the cascading event is necessary but might not be sufficient, to stop the damage from cascading.

6.6.2.2 Cascading Effects. Cascading effects occur when a hazard impacts a particular system and degrades its properties. Then for a

Figure 6-28. Cascading hazards during superstorm Sandy: high wind resulted in broken glass, which, in turn, resulted in rain damage to the store.
Source: Photo Courtesy of Mr. Albert Di Bernardo of Weidlinger Associates, Inc.

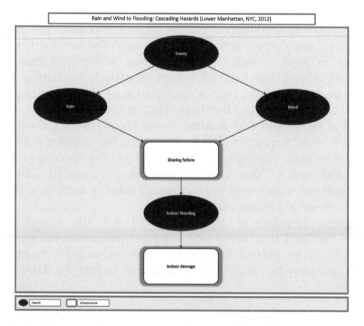

Figure 6-29. Cascading hazards, Lower Manhattan, NY, 2012.

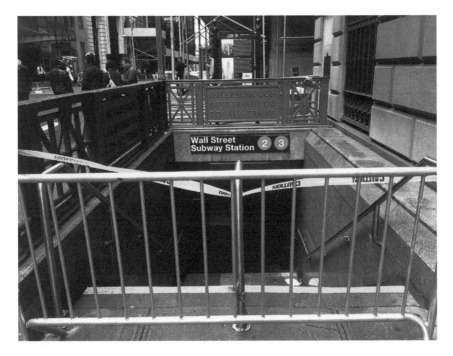

Figure 6-30. Cascading hazards during superstorm Sandy: storm surge led to the flooding of subway stations, which, in turn, led to an interruption of services. The resulting effects on numerous businesses through NYC were immense.
Source: Courtesy of Mr. Albert Di Bernardo of Weidlinger Associates, Inc.

completely unrelated reason, another hazard occurs, and then both hazards combine to create major damage that would not have happened if these two hazards did not combine. An example of the occurrence of cascading events can be traced during hurricane Sandy in the city of Long Beach, New York. While the strong wind that impacted the city was raging, a small fire accidentally started next to one of the residences. The fire was not caused by the strong wind; it was just an unfortunate coincidence. However, because of the strong wind, this minor fire became a raging fire that spread quickly and consumed several blocks of the city; see Figures 6-32 and 6-33.

Mitigation against cascading effects is more difficult than mitigation against cascading hazards. In this situation, mutual mitigation measures against both hazards are needed. In the City of Long Beach example, residences need protection not only against strong wind but also against fire. Most important, the possibility of accidents, especially during windstorms, triggered by the initiation of outside open fire, needs to be reduced or eliminated.

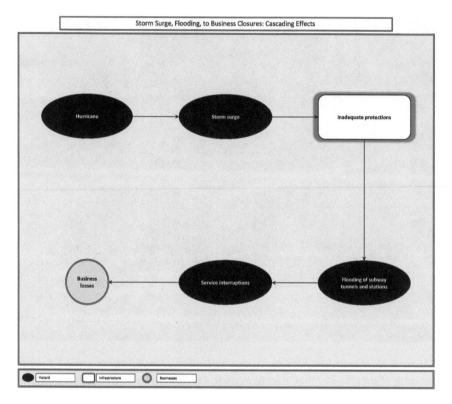

Figure 6-31. Cascading hazards, Hurricane Sandy, New York City Subway System, NY, 2012.

6.6.2.3 Cascading Failures. Cascading failures of an infrastructure, or a system of infrastructure, usually involve a single hazard, usually referred to as an initiating event. In such a case, the initiating event would cause a local failure in the system. This local failure would cause another failure, and a series of failures then continue to propagate throughout the system, causing considerable damage. Progressive collapse of buildings is a classic example; see Figure 6-34. Mitigation against cascading failures requires different sets of strategies than that against cascading hazards or cascading effects; see ASCE (2017a, b). A consideration of cascading failures (progressive collapse) is beyond the scope of this chapter.

6.6.3 Modeling of Cascading Events

The structure of networks that simulate cascading events of different types (events, hazards, failures, or combined modes) include the following:

1. Tree-structured network, see Figure 6-31.
2. Directional, noncyclic networks, see Figure 6-27, Figure 6-29, Figure 6-33, or Figure 6-35.

Figure 6-32. Cascading effects on private homes. Wind and minor fire spread quickly and resulted in damaging several city blocks.
Source: Courtesy of Mr. Marco Coco of Weidlinger Associates, Inc.

3. Directional and cyclic events, see Figure 6-34.
4. Combination of directional and nondirectional links (CGN).

A more extensive cascading effects model was offered by Ettouney and Alampalli (2017a). They modeled a flooding event that caused failure in a power station and how it can affect businesses and transportation infrastructure (mass transit stations and tunnels). In addition, the model showed the importance of redundancy considerations to arrest the cascading events. Ettouney (2022) offers a more complex CGN model for analyzing cascading hazards in the Siberian region of Russia. This model showcases a combination of directed and nondirected links among hazards as they cascade and a total of eight cascading hazards (Greenhouse Gases, GHGs; Global Warming, GW; Wildfires; Thawing of Permafrost; Sinkholes; Smoke; Insects; and Human Health).

6.6.4 Closing Remarks

A closer inspection of the three cascading events (hazards, effects, or failures) shows that they can mix in practice as Figure 6-35 illustrates symbolically, where H denotes a hazard and E denotes an event. Several

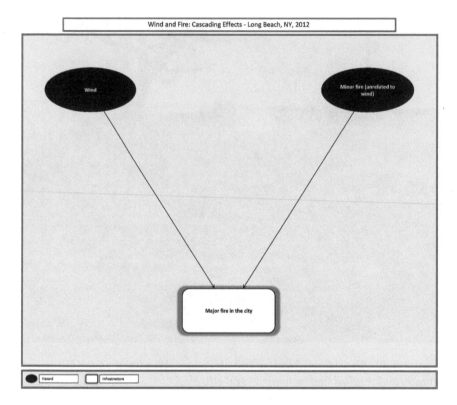

Figure 6-33. Cascading effects, Hurricane Sandy, City of Long Beach, NY, 2012.

possible paths for cascading events include H-1 → E-1 → E-8 → E-4 → E-7 → E-9. Some mixed cascading effects and cascading hazards are H-1 → E-3 → H-3 → E-6 → E-10. Note that even though modeling cascading events can be complex, network modeling, as explained in Section 6.2, can ease the demands of complexity. This shows, again, that a network modeling of resilience-related efforts is essential, given the important role that cascading behavior plays in any resilience consideration of an asset or community.

6.7 PREPAREDNESS

6.7.1 Overview

We used the Ostrich Paradox and its PIB corollaries in several applications so far in this chapter. This section, which is devoted to discussing issues related to preparedness, will offer a closer look at the subject. Recall that PIBs were offered by Meyer and Kunreuther (2017) as explanations of the impedances to public interest to preparedness. There is another model that offers explanations of constraints to communication between sources and destinations (following the Laswell communication model, see Section

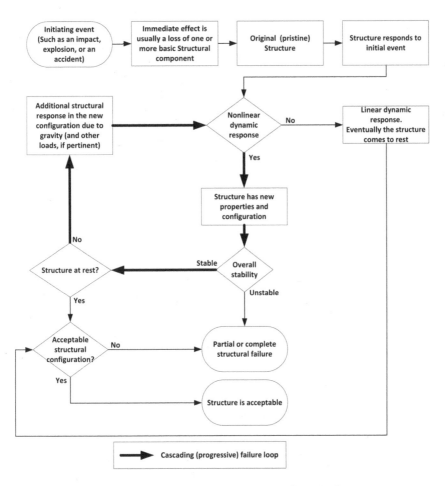

Figure 6-34. Cascading failure (AKA Progressive Collapse) of a structure.

6.3.6). This model was introduced by Lundgren and McMakin (2018) and is different in its context from the Ostrich Paradox, and because it applies to risk (or resilience) communication issues, we will refer to it as the Lundgren–McMakin constraints model (LMCM). The remainder of this section will explore the two models as they relate to preparedness and offer some objective processes to evaluate their applications.

6.7.2 Using Resilience Management Components to Address Preparedness Biases

A semiobjective approach to reduce PIBs would be to try to address them from the viewpoint of resilience management components. Tables 6-24 through 6-28 offer methods of addressing each of the PIBs from assessment, acceptance, treatment, monitoring, and communication perspectives.

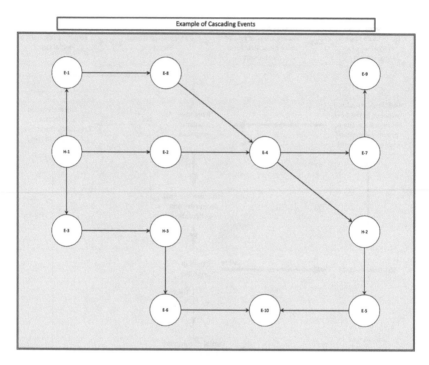

Figure 6-35. Subjective illustration of cascading effects and cascading hazards.

6.7.3 Utilizing Risk Management to Promote Preparedness

6.7.3.1 Behavioral Risk Auditing.   We explored back in Section 6.2.7 the Ostrich Paradox, see Meyer and Kunreuther (2017), which addressed the PIBs. We then offered BN models that aimed at estimating preparedness as a result of PIBs, Section 6.3.2.4.1, and PIBs as well as other important factors, Section 6.3.2.4.3. Of course, neither of these models addresses the important issue of how to control the issues that affect preparedness. Fortunately, Meyer and Kunreuther (2017) offered the BRA methodology that can reduce PIBs. Our objective now is to expand the aforementioned models to include, in an objective manner, the BRA process. While doing so, we will expand the BRA model to accommodate the additional factors beyond PIBs that also do affect preparedness.

On reflection, we observe that all of the techniques in the original BRA methods as offered by Meyer and Kunreuther (2017) can be categorized within one of the five components of risk management as we discussed them previously in the chapter. Because of this, we will use these five components as the starting variables in the model, and we will then recast the components of the BRA method within the model as needed. We will also introduce additional necessary components to the model for completeness.

Table 6-24. Methods to Reduce Preparedness Biases for Resilience Assessment.

Bias	Assessment
Myopia	Perform LCA-based assessments. Use DGN's Prediction in analysis.
Amnesia	Include past performances (using a DGN smoothing, that is, backward projections).
Optimism	Use reasonable confidence limits in analysis, so as to avoid overoptimistic results.
Inertia	Rely on up-to-date data for models. Allow for potential varied results.
Simplification	Insist on employing all pertinent underlying issues. Account for interactions through the use of network-based assessment.
Herding	Accommodate different models and/or their underlying input data for every problem.

Table 6-25. Methods to Reduce Preparedness Biases for Resilience Acceptance.

Bias	Acceptance
Myopia	Adjust acceptance threshold according to future forecasts.
Amnesia	Account for past experiences in setting current and future acceptance limits.
Optimism	Same as "Myopia".
Inertia	Change acceptance limits as needed. For example, allow for changing budgets or social needs in setting acceptance limits.
Simplification	Allow for complex acceptance models such as more than a single acceptance metric (budgets, subjective risk comfort, or reasonable time to recovery).
Herding	Allow for differing acceptance thresholds for different needs/moods of assets, communities, or regions.

Table 6-26. Methods to reduce Preparedness Biases for Resilience
Treatment/Improvement.

Bias	Treatment/improvement
Myopia	Institute policies requiring LCA considerations for public- and public-funded projects.
Amnesia	Institute policies that require past experiences, especially failures, to be included for all public- and public-funded projects. Encourage communication initiatives that emphasize past experiences as well as the advantages of incorporating them in current and future projects.
Optimism	Reasonable acceptance thresholds need to be used while designing new projects. Lowering these thresholds, for the sake of simplicity, might result in long-term additional costs.
Inertia	Encourage using new technologies and processes in projects.
Simplification	Consequences of executing a project need to be considered during the design and construction phases. Sometimes, unexpected consequences can lead to undesired results.
Herding	Allow for differences of settings among projects. No two project settings are 100% identical. Thus, recycling projects, although cost-effective, might not be advisable in the long run.

Note that the BRA method, and, thus, our model, is mainly applicable to short-term hazards, for example, floods, terrorism, and fire. We will address preparedness for long-term hazards in the next section. Also, we note that we use risk instead of resilience as a basis for our discussion because the original BRA process as introduced by Meyer and Kunreuther (2017) was risk-based, not resilience-based. Keeping the model risk-based would keep the spirit of the original model, without any loss of generality. Keep in mind also that, as we offered early on in this chapter that risk is a superset of resilience, so a risk-based model is fairly useful in the field of resilience management.

Table 6-27. Methods to Reduce Preparedness Biases for ReMo.

Bias	Monitoring
Myopia	Allow for monitoring systems to forecast future performances, in addition to monitoring the current states.
Amnesia	Use past-monitored data to aid in analyzing current data, as well as in forecasting future behavior.
Optimism	Strike a reasonable balance between underestimating and overestimating the results while analyzing monitored data.
Inertia	Encourage the use of new and advanced technologies in monitoring.
Simplification	Any monitoring system needs to be comprehensive so that it delivers its goals accurately. Trying to economize in designing a monitoring system might lead to erroneous conclusions.
Herding	Reasonable redundant and independent monitoring systems should be employed.

6.7.3.2 Modeling Risk Management and Behavioral Risk Auditing

6.7.3.2.1 Overview.   In this BN-based model, we try to model the interrelationship among

- Hazard (in the model we use flood hazard, following Meyer and Kunreuther (2017); even though the concept and the model can be easily adjusted to other short-term hazards);
- Different components of risk management, see Section 3;
- Different practical measures to reduce the PIBs; and
- PIBs themselves.

6.7.3.2.2 Model.   The variables in the model are described in Table 6-29 and pictured in Figure 6-36. Note that Figure 6-36 shows only abbreviated IDs for the variable names because of limited available space within the figure. The key to the full variable names can be found in Table 6-29. By preparing appropriate CPTs for different variables, the model can estimate states of preparedness as appropriate. Of course, different sensitivity studies can be performed with this model to answer different questions such as the following: (1) Given a desired preparedness state, what level of budget (or any other

Table 6-28. Methods to reduce Preparedness Biases for Resilience Communication.

Bias	Communication
Myopia	The communication media and message need to rely on forecasting results. Their complexities will also need to be consistent with the receivers (e.g., professionals, general public, and so on).
Amnesia	All communication components, see Gutteling and Wiegman (1996) or Lundgren and McMakin (2018), need to be consistent with past communication. They also need to highlight past experiences and their lessons for current and future needs.
Optimism	The communication message needs to be realistic. It needs to strike a proper balance between highlighting the hazard and its consequences and then between being prepared on the one hand and being misleadingly too optimistic on the other hand.
Inertia	There must be a reasonable balance between communication frequencies and its message and the temporal characteristics of the hazard it addresses.
Simplification	Communication plans need to focus on the communication message and media. The message needs to include all pertinent issues without oversimplification. The communication media need to be consistent with the receivers and the message itself.
Herding	Concentrate the communication plans on the receivers (public, regions, or commercial entities) and their varying specific needs.

controlling variable) is needed? Or (2) Given an observed level of a bias, for example, inertia bias, how would such an observation affect preparedness level?

We end this section by observing that the model of Table 6-29 and Figure 6-36, although complex, can be easily modified to accommodate any required special features or missing variables. Even though it is a BN model, some of the links can be changed so that they are nondirectional. The model can easily be changed into a decision model, or a dynamic model, if forecasting of some variables is needed.

Table 6-29. Variables of the BN Model for Risk Management, BRA,
and PIBs.

Variable ID in model	Full name of variables	Group	Comments
1	*Myopia*	PIBs	These are the original six biases that impede preparedness as introduced by Meyer and Kunreuther (2017).
2	*Amnesia*		
3	*Optimism*		
4	*Inertia*		
5	*Simplification*		
6	*Herding*		
7	*Preparedness Impedance Biases (PIB)*	Preparedness	Intermediate variable for preparedness. Metric for preparedness as a function of PIBs only.
8	*Hidden variable 1*	Hidden variables: introduced to improve computational efficiency	
9	*Hidden variable 2*		
10	*Infrastructure*	Parameters that control preparedness, in addition to PIBs.	Adequacy of different infrastructure preparedness levels.
11	*Budget*		Available budget to meet essential costs.
12	*Resources*		Measure of available resources to be adequately prepared: human, equipment, and so on.
13	*Other*		Any other variable that controls preparedness.
14	*Hidden variable 3*	Hidden variable: introduced to improve computational efficiency. This hidden parameter also shows an estimate of preparedness without PIBs.	

(Continued)

Table 6-29. (*Continued*) Variables of the BN Model for Risk Management,
BRA, and PIBs.

Variable ID in model	Full name of variables	Group	Comments
15	*Preparedness*	Preparedness	This variable represents the preparedness level as controlled by the different PIBs as well as other important variables.
16	*LCA*	Life Cycle Analysis	Can be used to reduce *Myopia* bias.
17	*CBA*	Cost–Benefit Analysis	Can be used to reduce *Optimism* bias.
18	*Assessment*	Risk management components	
19	*Acceptance*		
20	*Mitigation*		
21	*Monitoring*		
22	*Communication*		
23	*Insurance premium*	Risk mitigation	Transfer risk
24	*Insurance renewal*		
25	*Other mitigation measures*		Other mitigation measures can vary according to the region and level and type of hazard.
26	*Do nothing*		Doing nothing is an option in risk mitigation/ treatment.
27	*Improve infrastructure*		Hardening the physical infrastructure is always a popular option.

(*Continued*)

Table 6-29. (*Continued*) Variables of the BN Model for Risk Management,
BRA, and PIBs.

Variable ID in model	Full name of variables	Group	Comments
28	*Move*		Moving away from the hazardous region is an option in risk treatment.
29	*Laswell model*	Risk communication components. Note that this is not a comprehensive list.	The Laswell communication model includes five modules, see Figure 6-21.
30	*Message*		It includes messaging on the following: • Safety • Return on investment • Social responsibility
31	*Receivers*		• Simplify the message to accommodate communication's receivers as pertinent.
32	*Medium*		• Design the medium to suit the receivers and the message itself.

Note that the model does not include all of Laswell's communication components and does not include all possible mitigation measures.

We include only those variables that affect mainly the PIBs. For devising a more comprehensive model, additional variables will be needed.

6.7.4 Community Policies to Promote Preparedness

6.7.4.1 Overview.   The BRA approach that was developed by Meyer and Kunreuther (2017), which we reframed in a risk management model in Section 6.7.3, was meant to be applied for short-term hazards such as floods. Meyer and Kunreuther (2017) observed that it might not be applicable for long-term hazards such as sea-level rise (SLR). Instead, they offered a long-term solution for PIBs that is based on a set of four strategies that a community might adopt to prepare for SLR in the face of different

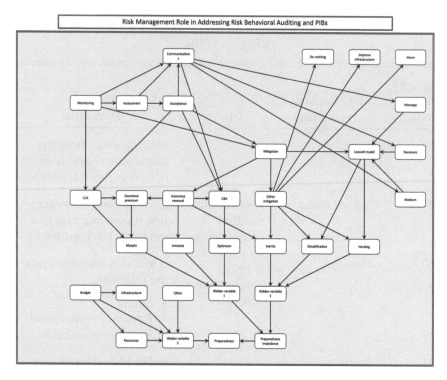

Figure 6-36. Risk management role in addressing BRA and PIBs.

PIBs that the community might be facing. These four strategies are as follows:

1. Enacting regulations, guidelines, and/or standards.
2. Changing or adapting zoning regulations.
3. Encouraging relocations to safer places, including a buyout program when pertinent.
4. Offering tax incentives (both short term and long term).

These strategies might be followed simultaneously, or perhaps might be followed with different degrees of administration, including not applying some of them at all. This, of course, will depend on the public reception in the community, as well as several other community-specific factors. The community administrators may choose how to apply these strategies subjectively, which may result in suboptimal results. Can we find an objective process that can help in achieving an optimal public reception to one, or a mix, of these strategies?

On reflection, it is apparent that we can cast this problem as a game of public indifference; see Section 6.2.6.4. One of the players would be the community administrators and the other is the public. Of course, we can add other players to the game such as financial concerns, developers, insurance companies, the state, and/or the federal government. However, we limit this section only to two players: community administrators and the public. We set the strategies of the community administrators to the aforementioned four strategies (ad developed by Meyer and Kunreuther 2017). The public sentiments (strategies) will be set to four, to simplify the solution, as follows:

1. Total acceptance.
2. Partial acceptance.
3. Partial rejection.
4. Total rejection.

Next, the community administrators will need to establish a payoff table for this 4 × 4 game (four promotional strategies by the administrators and four levels of public reception). The payoff table can be easily constructed via appropriate survey, personal experiences, historical records, or a combination of any of these methods.

The composition of the payoff table would control the solution method for this game. As before, see Section 6.3.6.1, we will consider two possible solution methods, the dominant equilibrium strategy and the Nash equilibrium mixed strategy.

6.7.4.2 Dominant Equilibrium Strategy.
Assume that the payoff table for this game is as in Table 6-30 and Figure 6-37. Looking for maxima of the community administrators for each of the four public sentiments, we find them to be 80, 65, 85, and 65, respectively, producing a dominant strategy of 85. Looking for the maxima (public sentiments) for each of the four community administrators' strategies, we find them to be 56, 60, 55, and 73, respectively, producing a dominant strategy of 73. We note that the third column and the fourth row contains the dominant strategies of both community administrators and the public, respectively. This means that there is a dominant strategy equilibrium of giving tax incentives by community administrators, which will produce a partial public rejection of the dominant strategy equilibrium. This is not a productive long-term strategy for the community administrators. A change, of course, would be needed. To improve public reception to the community strategies, the payoff, Table 6-30, needs to change. To facilitate these changes, the community needs to improve its preparedness promotional strategies; see Section 3.6.

6.7.4.3 Nash's Equilibrium Mixed Strategy.
Let us assume that after readjusting its preparedness promotional efforts, the community administrators researched and obtained new payoff values as in Table 6-31 and Figure 6-38. Looking for maxima of the community administrators for

Table 6-30. Payoff Table for Community Strategies to Promote Preparedness: Public indifference to a message with a Dominant Strategy Equilibrium.

Payoff for the strategies of the two players		Public sentiment							
		Total acceptance		Partial acceptance		Partial rejection		Total rejections	
		Payoff for the community	Degree of total acceptance	Payoff for the community	Degree of partial acceptance	Payoff for the community	Degree of partial rejection	Payoff for the community	Degree of total rejection
Strategies of the community to promote preparedness	Regulations/ guidelines/ standards	50	25	65	20	50	56	65	55
	Change/adapt zoning regulations	80	50	60	30	27	37	60	60
	Encourage relocations to safer places/ buyout when pertinent	50	45	65	55	50	23	65	5
	Tax incentives (both short term and long term)	80	15	60	30	85	73	60	41

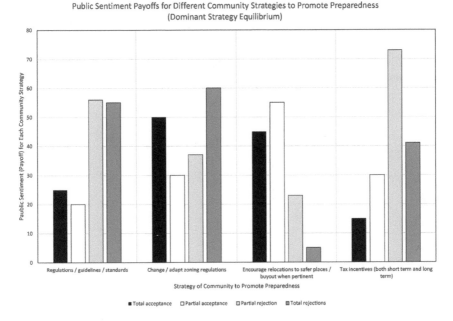

Figure 6-37. Payoff comparison for community strategies to promote preparedness: Public indifference to a message with a dominant strategy equilibrium.

each of the four public sentiments, we find them to be 80, 65, 85, and 65, respectively, producing a dominant strategy of 85 at the fourth row and third column. Looking for the maxima (public sentiments) for each of the four community administrators' strategies, we find them to be 56, 71, 55, and 41, respectively, producing a dominant strategy of 71 at the second row and third column. We note that the dominant strategies of both community administrators and the public do not coincide. This means that there is no dominant strategy equilibrium for this particular new payoff table. This means that for obtaining an optimal public sentiment, the community administration will need to use a Nash's equilibrium mixed strategy.

Turning back to the processes of the section "Public Indifference Game and Its Solution Methods," we can obtain the optimal strategy mix from solving Equations (6-16). Table 6-32 shows the resulting optimal mixed strategy. We find the probability mix of the four strategies that will promote public indifference to the messages, which is the optimal strategy mix in the long run.

6.7.4.4 Concluding Remarks. The examples in Sections 6.3.6.1.2, 6.3.6.1.3, 6.7.4.2, and 6.7.4.3 address public sentiment and how to mix the strategies

Table 6-31. Payoff Table for Community Strategies to Promote Preparedness: Public indifference to a message with Nash's Equilibrium and Mixed Strategy.

Payoff for the strategies of the two players		Public sentiment							
		Total acceptance		Partial acceptance		Partial rejection		Total rejections	
		Payoff for the community	Degree of total acceptance	Payoff for the community	Degree of partial acceptance	Payoff for the community	Degree of partial rejection	Payoff for the community	Degree of total rejection
Strategies of the community to promote preparedness	Regulations/ guidelines/ standards	50	25	65	20	50	56	65	55
	Change/ adapt zoning regulations	80	50	60	30	27	71	60	60
	Encourage relocations to safer places/ buyout when pertinent	50	45	65	55	50	23	65	5
	Tax incentives (both short term and long term)	80	15	60	30	85	10	60	41

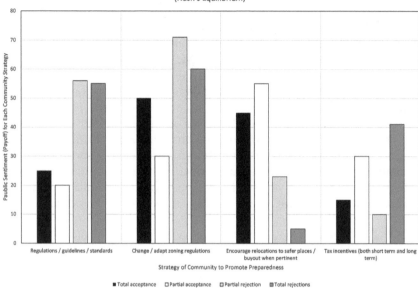

Public Sentiment Payoffs for Different Community Strategies to Promote Preparedness
(Nash's Equilibrium)

■ Total acceptance   □ Partial acceptance   □ Partial rejection   ▨ Total rejections

Figure 6-38. Payoff comparison for community strategies to promote preparedness: Public indifference to a message with Nash's equilibrium and mixed strategy.

Table 6-32. Strategy Mix for Promoting Preparedness.

Strategy	Probability mix
Regulations/guidelines/standards	0.48%
Change/adapt zoning regulations	38.28%
Encourage relocations to safer places/buyout when pertinent	30.14%
Tax incentives (both short term and long term)	31.10%

(policies in Decision Theory terminology) so as to optimize it. There were no considerations of costs (payoffs of the communication source or the strategy-/policy maker). Also, we considered only two players in these examples. Additional players such as insurance underwriters, financial institutions, and other local, state, and federal entities are needed to be included also as players. Such considerations can be accomplished via decision theory processes, but they are beyond the scope of this chapter.

6.7.5 Combining Ostrich Paradox, Lundgren and McMakin
Constraints Model, and Laswell Communication Models

We offered objective models that evaluate utilizations of PIBs in Sections 6.3.2.4, 6.7.3.2, and 6.7.4. While PIBs limit themselves to addressing preparedness issues only, Lundgren and McMakin (2018) offer a model that addresses communication constraints to the overall risk model. As such, both models, although different in scope and context, offer complementing and beneficial ways to address the preparedness issue from a wider perspective. In this section, we develop a simplified objective GN model that accounts for the general Lundgren and McMakin (2018) Constraints Model (LMCM) of risk communication in a Laswell framework, while accommodating the Ostrich Paradox principles. The objective of the model is to measure the effectiveness of risk communication, while considering the constraints of the LMCM and the PIBs of the Ostrich Paradox.

6.7.5.1 The Model. A GN with directed links that simulate Laswell communication flow is shown in Figure 6-39. The links among the nodes

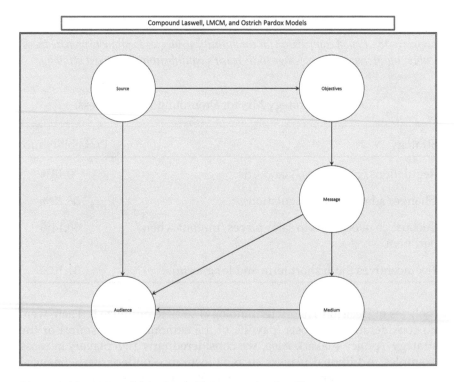

Figure 6-39. GN model for Laswell Communication Flow.

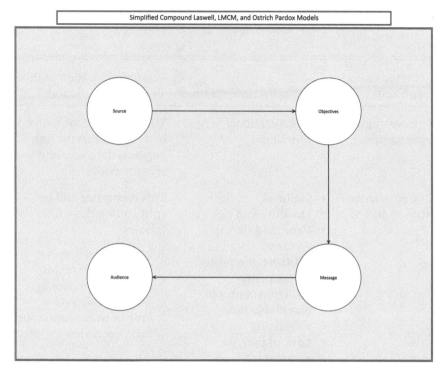

Figure 6-40. Simplified GN model for Laswell Communication Flow.

in the model will have impedances (weights) that will represent the strength of the constraints of the LMCM. To simplify the GN model even further, we use the model of Figure 6-40. This simplified model will join the *message* and *media* nodes. It will also combine Impedances (Constraints) #2 and #3 in Table 6-33 into a single constraint that is located between the new *message* and *audience* nodes. A closer examination of Constraint #2 in Table 6-33 reveals that it contains all of the PIBs, which implies that the GN of the Figures 6-39 and 6-40 model is actually a combination of the Ostrich Paradox, LMCM, and Laswell model.

We are now prepared to show how to use the model to evaluate the effectiveness of risk communication considering the LMCM. Let us assume that a particular organization is interested in evaluating the public reception of a communication message about a specific natural hazard. A study reveals that the three constraints of Table 6-33 are 20, 55, and 25 (weighted to total 100). Constraint #2 offers the highest impedance, so lowering it should improve the communication between the source and the audience. We note that the model as presented is extremely coarse.

Table 6-33. LMCM Constraints as Utilized in a Laswell Communication Model.

Communication constraints	Constrains descriptions	Assumed location within the Laswell model
1 - Constraints from source	• Organizational • Emotional	This constraint is assumed to be located on the link between the *source* and the *objective* nodes
2 - Constraints from audience	• Cultural • Hostility and outrage • Panic and denial • Apathy • Mistrust of reported assessments • Disagreements on acceptable risk thresholds • Lack of faith in science and institutions • Learning difficulties	This constraint will be split among three links as follows: 1. Link between *source* and *audience* nodes 2. Link between *message* and *audience* nodes 3. Link between m*edia* and *audience* nodes
3 - Constraints from interactions between source and audience	• Stigma • Stability of knowledge base	This constraint is assumed to be located on the link between the *source* and the *audience* nodes

Introducing more resolutions should improve its accuracy and benefits. For example:

- Introduce more detailed links to simulate specific impedances (constraints among the nodes.
- Add more nodes that simulate additional players (organizations, professionals, businesses, emergency managers, and so on) within the Laswell model.
- Explicitly simulate PIBs in the model.

The goal of developing such a detailed graph is to minimize the geodesic distances, $gd(source, audience)$, between the *source* and different *audience* nodes. Also, the diameter of such a graph, δ, should be minimized. See Section 2.4.1 for descriptions of the geodesic distance and diameter of graphs.

6.8 RETURN ON INVESTMENT

Resilience-related projects can bring great returns on investments (ROIs) for many stakeholders: public organizations (federal, state, and local); civil infrastructure owners; private industry (consulting engineers, vendors, constructors, and so on); as well as private residential property owners. There are many applications/opportunities in pursuing different resilience-related projects. So, while communities benefit from more resilient infrastructure, businesses and private citizens will also have an opportunity to increase their returns on investments (ROIs). For example, the Multihazard Mitigation Council (MMC 2005) showed that every $1.0 spent on mitigation projects can save $4.0. With this in mind, we discuss next, in a subjective manner, potential ROIs or sources of cost savings for different resilience management components.

6.8.1 Returns on Investment for Assessment Projects

A funder (a public organization that can be federal, state, or local), for example, might be interested in investing funds to improve the state of the economy. Let us assume that a proposed resilience-related project needs $M 0.5 in funding from the public organization. The project has two goals: providing accurate resilience assessment and then providing an accurate evaluation of a resilience improvement project. The project is in a region often affected by hurricanes, and there are 20 facilities in the region that are susceptible to the destructive force of hurricanes. The hurricane strength of interest is a 50 year hurricane. Such a hurricane is expected, on average, to cause losses per facility per event estimated at $2 million. These losses include disruptions in operations, physical damage, and other indirect losses. Such losses can be reduced by about 5% if appropriate resilience enhancement measures (learned from an accurate resilience assessment project) are enacted. These savings are realized by slight modifications in operating practices without any major capital expenditures. Based on this, we can compute an expected cumulative rate of return on investment (ROI), as shown in Figure 6-41. The ROI after 5 years on resilience-related projects from a public organization's viewpoint is considerable.

This example is based on intentionally conservative assumptions of (1) assumed savings from identifying weak points in the facilities, (2) exposure to only one hurricane, and (3) assumed savings from optimal project evaluations. Perhaps the most conservative assumption is the expected loss per event. Figure 6-1 shows that losses from major events can be much higher than those resulting in our example. More realistic assumptions would produce higher ROI. However, the example shows that even with extremely conservative assumptions, the ROIs that public organizations can realize from funding resilience-related projects can be considerable.

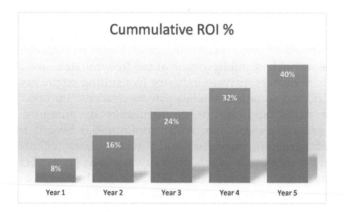

Figure 6-41. Example of cumulative ROI on a publicly funded resilience project.
Source: Ettouney (2014).

6.8.2 Return on Investments for Resilience Acceptance Projects

Resilience acceptance thresholds can have enormous cost implications. The total cost of any event or entity (asset or community) that might require resilience management is composed of two parts: the total cost of the damage caused by the event itself (including all direct and indirect costs) and the cost of retrofits that might improve resilience. It is obvious that as the cost of a retrofit increases, resilience will increase and damage costs will decrease. The reverse is also true. These relationships are shown in Figure 6-42, which also shows the total cost (the sum of both damage

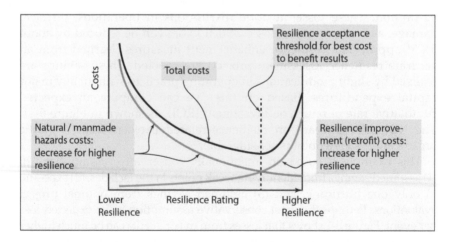

Figure 6-42. Business case for resilience acceptance.
Source: Ettouney (2014).

costs and retrofit costs). The total cost will have a "sweet spot" resilience point where total costs can be kept to a minimum. If the resilience acceptance threshold moves away, in either direction, from this "sweet spot" threshold, the total cost will increase. Thus, any decision maker should strive for a resilience acceptance threshold that is as close to this "sweet spot" threshold as possible. This can be achieved only by careful estimations of damage costs as well as retrofit costs.

We observe that Figure 6-42 shows only one level of a hazard demand. Other relationships can be made for different demand levels. The resilience "sweet spot" acceptance limit is expected to differ for those different demands. In addition to demand levels, several other factors can impact the acceptance "sweet spots." For example, 1 = lifecycle cost/benefit, 2 = multihazard considerations, and 3 = potential of cascading effects.

6.8.3 Return on Investments for Resilience Improvement Projects

Decision makers need to execute two steps when considering resilience improvement projects: Step 1: Identify several potential projects, then prioritize the projects according to an appropriate cost/benefit as well as LCA criteria. Note that costs versus benefits need not be only monetary in nature. Step 2: Given the available project list developed in Step 1, consider budgetary availability and implications during the process of selecting a project or several projects, as needed.

Finding the ROI for cost/benefit ratios in an LCA framework of a particular project is not an easy task. FEMA (1996), MMC (2005), and Multihazard Mitigation Council (MMC 2017) have studied this issue in depth. MMC (2005) concluded that, on average, the benefit from natural hazard mitigation projects is about $4.0 for every $1.0 cost. This 400% ROI is a large value that should be taken seriously. Even on an annual basis, if we assume, for example, that, on average, the life span of mitigation projects is approximately 25 years, we end up with an average ROI of about 16%. Again, this ROI value is higher than any reasonable annual discount rate. As a comparison point, the rate of returns of S&P 500 since its inception averaged 10% to 11%; see Maverick (2020). Given that the 16% ROI is only an average, it can be improved by following an objective prioritization process. Based on this, we reach an important conclusion: entities that develop and follow objective resilience-improvement prioritization processes can improve the average annual RIO from 16% to a much higher number.

6.8.4 Return on Investments for Resilience Monitoring Projects

Let us consider a situation in which an owner has a facility with an acceptable resilience threshold (because at this threshold, the owner's studies predict an acceptable cost for a particular hazard level or multihazard levels as pertinent). In addition, the facility owner estimates that the accepted threshold will not be crossed because of changes in demands, capacity, social, and economic reasons before a certain time in the future. A capital retrofit plan was prepared for execution at that future estimated time.

If, because of some unpredictable future events, the actual future system capacity, demands, or consequences exceed (in case of demands or consequences), or fall below (in case of capacity), their respective estimated values, then the accepted resilience threshold will be reached faster than predicted by the resilience study. Thus, if the facility owner continues with the original capital expenditure plan, the actual future system resilience level will be less than the estimated one.

A cost-effective approach would be to adopt a resilience monitoring (ReMo) plan that will show, in real time, or near real time, the instant when the acceptable resilience threshold is reached. Adequate modifications in the original plans could then be made to avoid increase in the cost below par resilience state. It is reasonable to assume that the ROI of ReMo will prove to be attractive when compared with the cost of a subpar resilience. Figure 6-43 illustrates the potential of the ReMo value.

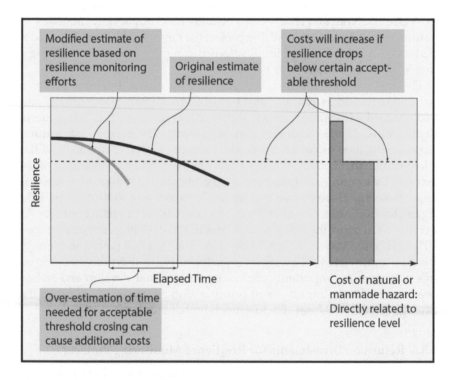

Figure 6-43. Business case for ReMo.
Source: Adapted from Ettouney (2014).

Objective methods that estimate degradation of systems as time passes are well documented; see, for example, Agrawal and Kawaguchi (2009). Ettouney and Alampalli (2011a, 2011b) discussed the value of monitoring and how to estimate their ROI.

6.9 SUMMARY AND CONCLUSIONS

With increased costs of man-made and natural hazards, cost-effective methods are needed to provide for resilient systems (assets and communities). A key to realizing optimal resilience considerations, objective processing is needed. This chapter started with an exploration of resilience definitions, asset versus community needs, comparisons of the concepts of risk, resilience, and sustainability, and the multidimensionality of resilience. We then offered several objective methods that are suitable for addressing the subject of the resilience of assets and communities. This included the theory of graph networks, probabilistic graph networks, and game theory. We then discussed the different dimensionalities of resilience that included multidisciplinary (interagency) efforts, multihazard interactions, cascading considerations, and resilience management. We argued that a comprehensive resilient management approach can achieve cost-efficient resilient systems. The chapter explored different aspects of resilience management for civil infrastructure, including assessment, acceptance, treatment, monitoring, and communication. Several practical examples were presented.

Our final remarks in the chapter offered a subjective discussion regarding potential ROIs for specific components of resilience management. Based on the discussion, it is evident that stakeholders of civil infrastructure stand to benefit from accounting for all resilience dimensions as well as resilience management to reach the desired goal of resilient systems in an efficient and cost-effective manner.

6.10 RECOMMENDED PRACTICES

Based on the discussions of this chapter, we can make some recommended practices, as follows:

1. Resilience considerations scale up from single assets to whole communities. To achieve acceptable community resilience considerations, the resilience of its component assets must be accommodated.
2. Resilience is a multidimensional subject. Important resilience dimensions are resilience management, multihazard considerations, multidisciplinary considerations, and cascading events. All these dimensions must be accounted for.

3. Even though resilience can be expressed subjectively and objectively, only objective considerations can assure optimal resilient systems.
4. Interactions (sometimes referred to as links) among different factors that control dimensions of resilience, see # 2, need to be accommodated.
5. Components of resilience management—assessment, acceptance, treatment, monitoring, and communication—are integral parts of optimal resilient systems. Resilience assessment is a necessary but far from sufficient aspect of resilience.
6. Careful considerations of the subtle differences among cascading events (cascading hazards, cascading effects, and cascading failures) are needed while managing the resilience of any system.
7. Resilience of a system can be managed only in the context of a hazard or a group of hazards. In case of a multitude of potential hazards, multihazard considerations need to be managed.
8. Resilient systems require multidisciplinary (interagency) cooperation. There cannot be adequate resilient assets or communities without effective multidisciplinary underpinnings.
9. Building resilient systems is an investment. Similar to any other investments, returns on this investment might not be immediate, but in the context of a life cycle, the ROI of resilience investment will offer high returns.

NOTATIONS

BN	Bayes network
BRA	Behavioral risk auditing
CE	Cascading events
DCG	Dynamic chain graph
DCGN	Dynamic chain graph network
DPGN	Dynamic probabilistic network
GN	Graph network(s)
GT	Game theory
ID	Influence diagram
LMCM	Lundgren–McMakin constraints model
MDT	Multihazard decision theory
MH	Multihazards
MN	Markov network (sometimes referred to as Markov random fields)
MOT	Multihazard operations theory
MPT	Multihazard Physical Theory
MRR	Maintenance, repair, rehabilitation.
PGN	Probabilistic graph network(s)

PIB   Preparedness impedance bias
ReMo   Resilience monitoring
ROI   Return on investment
RTS   Rapid transit systems
SLR   Sea-level rise
SOP   Standard operating procedures
TLN   Time link node

REFERENCES

Agrawal, A., and A. Kawaguchi. 2009. *Bridge element deterioration rates.* Albany, NY: New York State Department of Transportation.

Alampalli, S., and M. M. Ettouney. 2010. "Resiliency of bridges: A decision making tool." *Bridge Struct.: Assess. Des. Constr.* 6 (1–2): 8.

Alampalli, S., and M. Ettouney. 2016. "Suspension bridge security risk management." In *Inspection, evaluation and maintenance of suspension bridges,* edited by S. Alampalli and W. Moreau, 301–346. Boca Raton, FL: CRC Press.

Annandale, G. 2006. *Scour technology.* New York: McGraw Hill.

ASCE. 2013. *2013 Report card for America's infrastructure.* Reston, VA: ASCE.

ASCE. 2017a. *Minimum design loads and associated criteria for buildings and other structures—Commentary.* Reston, VA: ASCE.

ASCE. 2017b. *Minimum design loads and associated criteria for buildings and other structures—Provisions.* Reston, VA: ASCE.

Betti, R. 2022. "Monitoring for resilience in highway bridges." Chap. 3 in *Objective resilience: Technology,* edited by M. M. Ettouney, MOP 148, 53–74. Reston, VA: ASCE.

Bruneau, M., S. Chang, R. Eguchi, T. O'Rourke, A. Reinhorn, M. Shinozuka, et al. 2003. "A framework to qualitatively assess and enhance the seismic resilience of communities." *Earthquake Spectra* 19 (4): 20.

Bruneau, M., and A. Reinhorn. 2007. "Exploring the concept of seismic resilience for acute care facilities." *Earthquake Eng. Res. Inst. Earthquake Spectra* 23 (1): 41–62.

Chavel, B., and J. Yadlosky. 2011. *Framework for improving resilience of bridge design.* Washington, DC: Federal Highway Administration.

Deo, N. 1974. *Graph theory with applications to engineering & computer science.* Mineola, NY: Dover.

DHS (Department of Homeland Security). 2009. *Costs and benefits of natural hazard mitigation.* Washington, DC: DHS.

DHSEM (Division of Homeland Security and Emergency Management). 2019. *Emergency management office organization chart.* Centennial, CO: DHSEM.

Ettouney, M. M. 2014. *Resilience management: How it is becoming essential to civil infrastructure recovery*. New York City: McGraw Hill.

Ettouney, M. M. 2022. "Climate resilience." Chap. 6 in *Objective resilience: Applications*, edited by M. M. Ettouney, MOP 149, 189–310. Reston, VA: ASCE.

Ettouney, M., S. Alampalli, and A. Agrawal. 2005. "Theory of multihazards for bridge structures." *Bridge Struct.* 1 (3): 11.

Ettouney, M. M., and S. Alampalli. 2011a. Vol. 2 of *Infrastructure health in civil engineering: Applications and management*. Boca Raton, FL: CRC Press.

Ettouney, M. M., and S. Alampalli. 2011b. Vol. 1 of *Infrastructure health in civil engineering: Theory and components*. Boca Raton, FL: CRC Press.

Ettouney, M. M., and S. Alampalli. 2017a. *Multihazard considerations in civil infrastructure*. Boca Raton, FL: CRC Press.

Ettouney, M. M., and S. Alampalli. 2017b. *Risk management in civil infrastructure*. Boca Raton, FL: CRC Press.

FEMA. 1996. *Costs and benefits of natural hazard mitigation*. Washington, DC: FEMA.

FEMA. 2005. "Risk assessment: A how-to guide to mitigate terrorist attacks." In *Risk management series*, edited by M. Kennett. Washington, DC: FEMA.

FEMA. 2020. *Avoiding wildfire damage: A checklist for homeowners*. Washington, DC: FEMA.

Fenton, N., and M. Neil. 2013. *Risk assessment and decision analysis with Bayesian networks*. Boca Raton, FL: CRC Press.

Flanigan, K. A., M. Aguero, R. Nasimi, F. Moreu, J. P. Lynch, and M. M. Ettouney. 2022. "Objective resilience monitoring for railroad systems." Chap. 4 in *Objective resilience: Technology*, edited by M. M. Ettouney, MOP 148, 75–120. Reston, VA: ASCE.

Fudenberg, D. 1991. *Game theory*. Cambridge, MA: MIT Press.

Gerasimidis, S., and M. Ettouney. 2022. "On the definition of resilience." Chap. 1 in *Objective resilience: Policies and strategies*, edited by M. M. Ettouney, MOP 146, 1–24. Reston, VA: ASCE.

Gibbons, R. 1992. *Game theory for applied economists*. Princeton, NJ: Princeton University Press.

Gromicko, N. 2006. "Wildfire mitigation strategies and inspection." Accessed September 8, 2012. https://www.nachi.org/wildfire-mitigation-strategies-inspection.htm.

Gutteling, J., and O. Wiegman. 1996. *Exploring risk communications*. Dordrecht, Netherlands: Kluwer.

Hughes, J., and K. Healy. 2014. *Measuring the resilience of transport infrastructure*. Wellington, New Zealand: NZ Transport Agency.

Hughes, S. 2022. "Resilience of rapid transit systems: A practical outlook." Chap. 5 in *Objective resilience: Applications*, edited by M. M. Ettouney, MOP 149, 101–188. Reston, VA: ASCE.

Joyner, D., M. V. Nguyen, and N. Cohen. 2012. "Algorithmic graph theory." Version 3.0. Accessed at May 10, 2013. http://code.google.com/p/graphbook/.

Kennett, M., M. Ettouney, S. Hughes, R. F. Walker, and E. Letvin. 2011a. "Integrated rapid visual screening of mass transit stations." In *Buildings and infrastructure protection series*, edited by M. Kennett. Washington, DC: Dept. of Homeland Security.

Kennett, M., M. Ettouney, S. Hughes, R. F. Walker, and E. Letvin. 2011b. "Integrated rapid visual screening of tunnels." In *Buildings and infrastructure protection series*, edited by M. Kennett. Washington, DC: Dept. of Homeland Security.

Kennett, M., M. Ettouney, P. Schneider, R. F. Walker, and M. Chipley. 2011c. "Integrated rapid visual screening of buildings." In *Buildings and infrastructure protection series*, edited by M. Kennett. Washington, DC: Dept. of Homeland Security.

Koller, D., and N. Friedman. 2009. *Probabilistic graphical models principles and techniques*. Cambridge, MA: MIT Press.

Lasswell, H. 1948. *The communications of ideas*, edited by L. Bryson, 37–51. New York: Harper.

Lundgren, R., and A. McMakin. 2018. *Risk communication: A handbook for communicating environmental, safety, and health risks*. Hoboken, NJ: Wiley.

Maverick, B. 2020. "What is the average annual return for the S&P 500?." Accessed September 22, 2020. https://www.investopedia.com/ask/answers/042415/what-average-annual-return-sp-500.asp.

Mertz, D. 2012. *Steel bridge design handbook: Redundancy*. Washington, DC: Federal Highway Administration.

Meyer, R., and H. Kunreuther. 2017. *The ostrich paradox*. Philadelphia: Wharton School Press.

MMC (Multihazards Mitigation Council). 2005. *Natural hazard mitigation saves: An independent study to assess the future savings from mitigation activities*. Washington, DC: National Institute of Building Sciences.

MMC. 2017. *Natural hazard mitigation saves: 2017 Interim report*. Washington, DC: National Institute of Building Sciences.

Mohammadfam, I., S. Bastani, M. Esaghi, R. Golmohamadi, and A. Saee. 2014. "Evaluation of coordination of emergency response team through the social network analysis. Case study: Oil and gas refinery." *Safety Health Work* 6 (1): 30–34.

NCEI (National Centers for Environmental Information). 2020. "U.S. billion-dollar weather and climate disasters." Accessed September 4, 2020. https://data.nodc.noaa.gov/cgi-bin/gfx?id=gov.noaa.nodc:0209268.

Neapolitan, R. E. 2004. *Learning Bayesian networks*. Upper Saddle River, NJ: Prentice-Hall.

Newman, M. E. J. 2010. *Networks: An introduction*. Oxford, UK: Oxford University Press.

NIAC (National Infrastructure Advisory Council). 2009. *Critical infrastructure resilience final report and recommendations*. Washington, DC: NIAC.

NRC (National Research Council). 2010. *Review of the Department of Homeland Security's approach to risk analysis*. Washington, DC: National Academies Press.

NSC (National Security Council). 2011. "Presidential Policy Directive/PPD-8: National preparedness." Accessed May 26, 2018. https://www.dhs.gov/presidential-policy-directive-8-national-preparedness.

Office of the Press Secretary. 2013. "Presidential Policy Directive/PPD-21: Critical infrastructure security and resilience." Accessed October 20, 2019. https://www.dhs.gov/sites/default/files/publications/PPD-21-Critical-Infrastructure-and-Resilience-508.pdf.

Osborne, M., and A. Rubinstein. 1994. *A course in game theory*. Cambridge, MA: MIT Press.

Powell, W. 2011. *Approximate dynamic programming: Solving the curses of dimensionality*. Hoboken, NJ: Wiley.

Prisner, E. 2014. *Game theory through examples*. Washington, DC: Mathematical Association of America.

Spaniel, W. 2015. *Game theory 101: The complete textbook*. Middletown, DE: William Spaniel.

Tehrani, F. M., and D. Nelson. 2022. "From sustainability to resilience: A practical guide to ENVISION." Chap. 3 in *Objective resilience: Objective process*, edited by M. M. Ettouney, MOP 147, 81–126. Reston, VA: ASCE.

Vose, D. 2009. *Risk analysis: A quantitative guide*. Hoboken, NJ: Wiley.

WEF (World Economic Forum). 2017. *The global risks report 2017*. Geneva: WEF.

INDEX

Note: Page numbers followed by *f* and *t* indicate figures and tables.